Table of Contents

Preface

Frankho ChessDoku™ puzzles were invented by Mr. Frank Ho, a Canadian math teacher and the founder of Ho Math Chess Tutor Franchise Learning Centre. The puzzles included in this workbook were jointly created by Frank Ho and Amanda Ho.

Game play is a good way to inspire children's interest in learning math and chess has proven to be effective as a learning tool. Frankho ChessDoku™ is a collection of a special kind of math puzzles that are solved by using addition, subtraction, multiplication, division, or number factoring by following chess moves, logic, and backwards thinking strategy. Frankho ChessDoku™ is one-of-a-kind puzzle that can help children improve their computing, logic, and chess abilities all in one workbook and at the same time.

Since most children like puzzles and games, so for children working on Frankho ChessDoku™ is a good alternative for them to learn math computation skills whether they are interested in playing chess or not. There are no chess knowledge required at all when working on Frankho ChessDoku™ because children could simply follow the directions as indicated by darker lines. The numbers involved in the computations are all just one-digit numbers.

Frankho ChessDoku™ puzzles were created by using the following Frank's invented and trademarked Geometric Chess Symbols which are also used for Ho Math Chess Teaching set. A child as young as four years old can play chess almost instantly by learning this chess set.

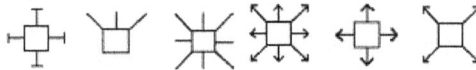

(Canada copyright 1069744, trademark TMA 771400)

When working on Frankho ChessDoku™ puzzles, children explore the calculation pathways by using clues such as common squares intersected by chess moves. This logical thinking process adds a fun element to the learning of basic number facts.

What makes Frankho ChessDoku™ intriguing is that it teaches children logic and the decision making process. Even though there is only one final answer; the immediate answers in the process of calculating may be more than one. Many math concepts such as operations of intersections and reversing order of operations such as finding addends of a given sum are often included in the thinking process. Children also are taught the concepts of line interactions (chess moves) and logic while having fun following chess moves.

Frankho ChessDoku™ puzzles are educational, fun, and addictive.

For more details, please contact Ho Math Chess at mathandchess@telus.net.

Frank Ho
Amanda Ho

October, 2008

Student's name _____ Date _____

What are specials about Frankho ChessDoku™ puzzles?

- They are unique math puzzles (over 200) combining with math, chess, Sudoku, and logic.

- They do not require any high-level math skills other than addition, subtraction, multiplication, division, factoring, comparisons, and visualization.

- The chess skills required are very elementary. Students are only required to know how each chess piece moves and how it is represented by a corresponding Geometric Chess Symbol – invented and trademarked by Frank Ho.

- All the puzzles can be solved by students with minimum teacher's guidance.

- The puzzles train students to be patient and use their problem-solving ability. They are fun and addictive.

- The abilities learned by doing the puzzles can be applied to help students succeed in their day school math courses.

Ho **Math Chess = A Cool and Fun Way to Learn Math!** ™

Why children like Frankho ChessDoku™?

One question puzzled myself when I observed that our children's likeness of Frankho ChessDoku exceeding our expectation. Why? I started to look into the reasons and also did a little experiment to verify if our children really like Frankho ChessDoku and the following is my report.

Frankho ChessDoku is really a great invention for being used as a fun supplemental basic number facts training material. We have tested this product in our own Vancouver learning centre since year 2008 and discovered Frankho ChessDoku 3 by 3 has been always been the most favorite workbook for primary students and this can be testified by students when they were asked to choose the most liked workbook out of their workbooks. Some parents even told me that their kids would voluntarily to work on Frankho ChessDoku on their own without being asked to do so,

Why our students like Frankho ChessDoku workbook? What is the secret?

1. It benefits children and they know it. Not only Frankho ChessDoku improves a child's basic number facts computation ability, it can also improves their logic thinking and visualization skills and even chess play ability because children are trained to watch for the intersections of chess moves while working on Frankho ChessDoku .

2. Frankho ChessDoku is not boring because every problem is different. The problem presents a mystery to children and make them want to solve it simply it is not too difficult yet it is also challenging for them and the chess moves encourages them to solve the puzzle one step at a time.

3. There are no lengthy English sentences to comprehend before they could solve the problem. Many times they face a multiple answers but not sure which one is right so they have to put on their logic thinking hat to try to identify the correct answer. While they are working on arithmetic problems, it is not straight to find the answer 2 + 3, instead it is what numbers would make 3 by using some kind of arithmetic operations, so that is why they do not feel bored.

4. Often children feel bored when working on word problems because most of the time when they get stuck then they feel hopeless to continue on, but this kind of feeling does not exist in Frankho ChessDoku because they could tackle the same problem form different directions by following chess moves and try to find the answer.

5. Often children feel bored when working on pure computation worksheets because they constantly have to do arithmetic sheets after sheets and they do not like those multiple-digit numbers. With Frankho ChessDoku, children only using numbers 1, 2, 3, so the calculations is not so difficult and even a kindergarten could work on Frankho ChessDoku problems.

6. The final result always follow Sudoku rule so children could check on their own answers to see if the result is correct or not. The entire learning process also trains children to be a self-learner form the beginning to the end. Teacher's or parent's guidance is kept to minimal.

7. After finishing Frankho ChessDoku, there is no massy room to be cleaned up and no scattered manipulative to be collected, students get good brain workout by only using a pencil and an eraser. Both teachers and children feel happy in this kind of organized environment.

8. Students could even help each other to have a good team spirit when working on Frankho ChessDoku.

9. Even though Frankho ChessDoku involves chess, it absolutely does not require students to know how to play chess because students simply follow the directions of lines.

How to solve Frankho ChessDoku ™

Rule: The numbers 1, 2, 3 must appear only once in every row or column of the following ChessDoku problem.

Step 1:   The squares with numbers in them shall be the ones we pay attention first. The math strategy is to work backwards. These numbers are results of calculations according to some arithmetic operator(s) and chess moves(s) as indicated by darker arrow(s).

Step 2:   First let's look at number 5 in the c3 square. The 5 is a result of chess piece Queen's move coming from the left, so let's think what two numbers add to 5? The answer is 2 + 3 or 3 + 2 but we are not sure which one to choose, so we now take a look at the number 6 in b2 square.

Step 3:   The number 6 is the result of 3 + 3 and no other combination of two numbers (from the number choices of 1, 2, and 3) can be added to have the result of 6. We then can conclude that the number 3 which we could not decide at the Step 2 shall be in the c3 square.

Step 4:   Now we have three squares figured out, they are 2 in b3, 3 in c3, and 3 in b2. By using the strategy of the rules of Sudoku (each number only appears once per row or column.) we know 1 must be placed in square a3 and also 1 must be placed in square b1.

Step 5:   Now we look at the square which is intersected by 2 other squares having numbers written in them. For example, square c1 is intersected by square b1 (1) and c3 (3) (Note in this case 3 points b1, c1, and c3 forming a triangle so we can call it Triangle strategy). We know if the number 1 already appeared in row 1 and the number 3 already appeared in column c, then the square c3 must have the number 2. Continue to use the restriction of no same numbers can be appeared on the same row and column, we can figure out all other numbers.

To watch Frank Ho's video presentation explaining the following problem, go to
https://www.youtube.com/watch?v=CgP6ZnTYTSk&feature=youtu.be&hd=1

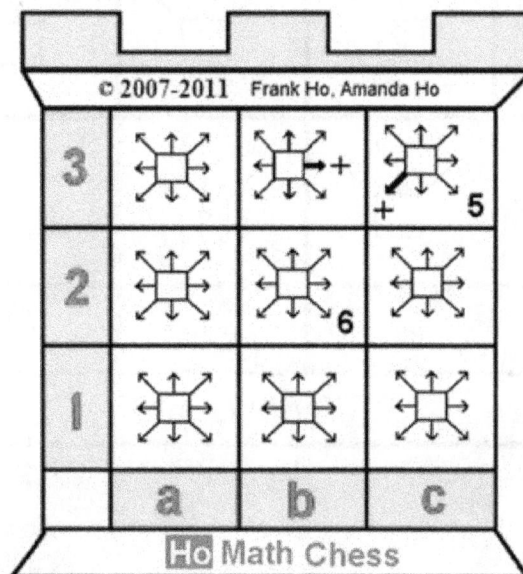

***** Part 1 Frankho ChessDoku 3 by 3 *****

Frankho ChessDoku™ # 1

Frankho ChessDoku™ is solved by using one or more operators of addition, subtraction, multiplication, or division after following chess moves and logic.

Rule All the digits 1 to 3 must appear exactly once in every row and column. The number appears in the bottom right-hand corner is the end result calculated according to arithmetic operator(s) and chess move(s) as indicated by darker arrow(s).

Pawn

The following figure shows how pawn moves by following its darker line segment(s).

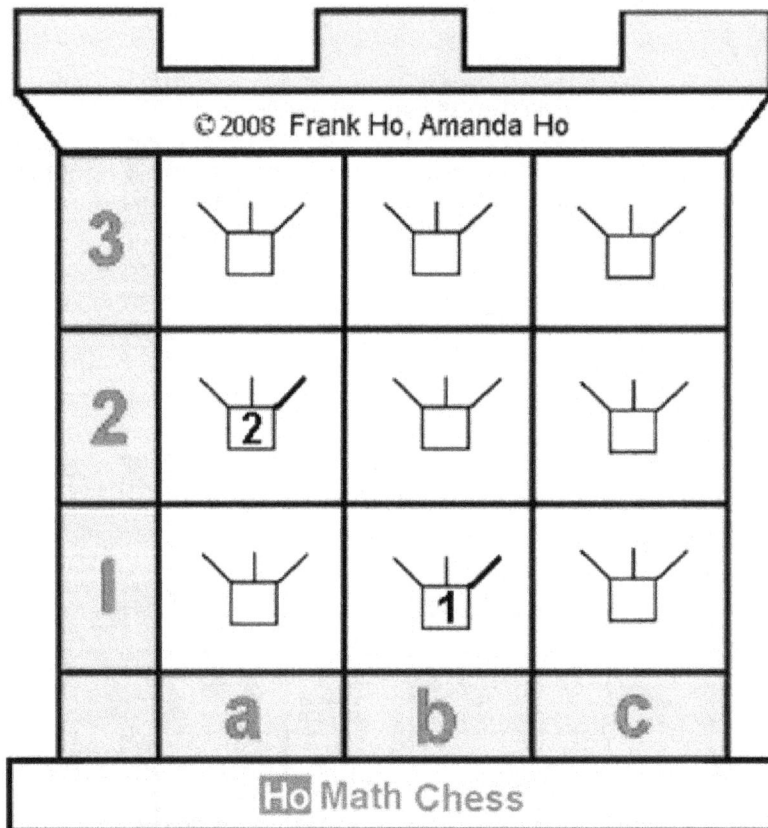

Student's name _____   Date _____

Frankho ChessDoku™ # 2

Frankho ChessDoku™ is solved by using one or more operators of addition, subtraction, multiplication, or division after following chess moves and logic.

Rule All the digits 1 to 3 must appear exactly once in every row and column. The number appears in the bottom right-hand corner is the end result calculated according to arithmetic operator(s) and chess move(s) as indicated by darker arrow(s).

Bishop

The following figure shows how bishop moves by following its darker line segment(s).

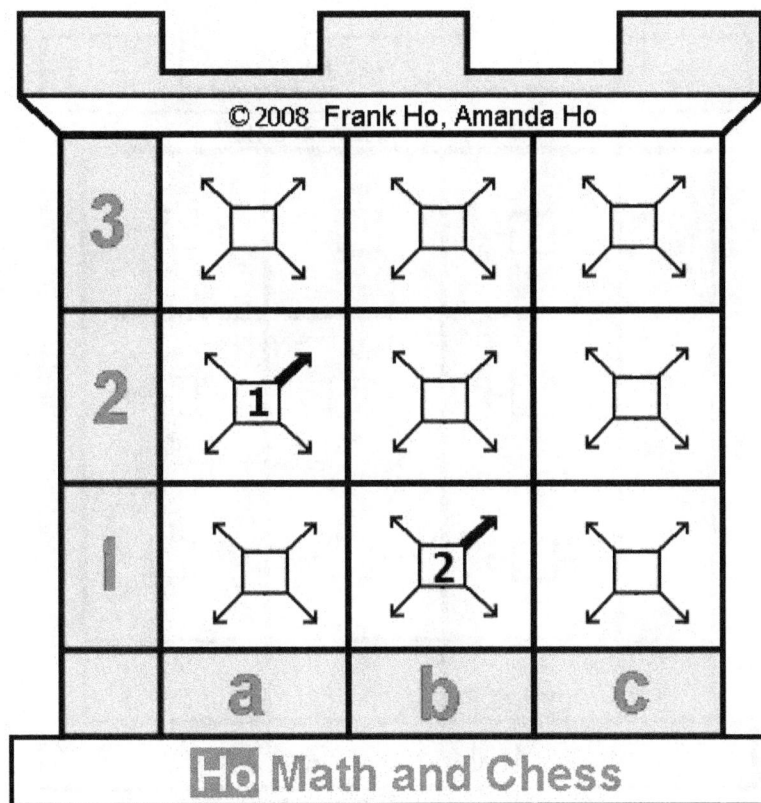

© 2008 Frank Ho, Amanda Ho

Ho Math and Chess

Frankho ChessDoku™ # 3

Frankho ChessDoku™ is solved by using one or more operators of addition, subtraction, multiplication, or division after following chess moves and logic.

Rule All the digits 1 to 3 must appear exactly once in every row and column. The number appears in the bottom right-hand corner is the end result calculated according to arithmetic operator(s) and chess move(s) as indicated by darker arrow(s).

Rook

The following figure shows how rook moves by following its darker line segment(s).

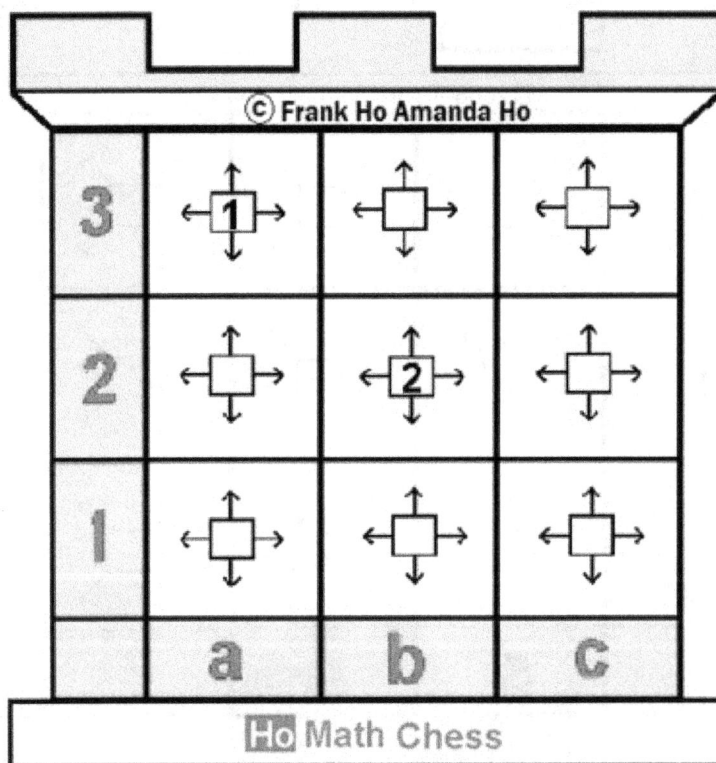

© Frank Ho Amanda Ho

Ho Math Chess

Frankho ChessDoku™ # 4

Frankho ChessDoku™ is solved by using one or more operators of addition, subtraction, multiplication, or division after following chess moves and logic.

Rule All the digits 1 to 3 must appear exactly once in every row and column. The number appears in the bottom right-hand corner is the end result calculated according to arithmetic operator(s) and chess move(s) as indicated by darker arrow(s).

Knight

The following figure shows how knight moves by following its darker line segment(s).

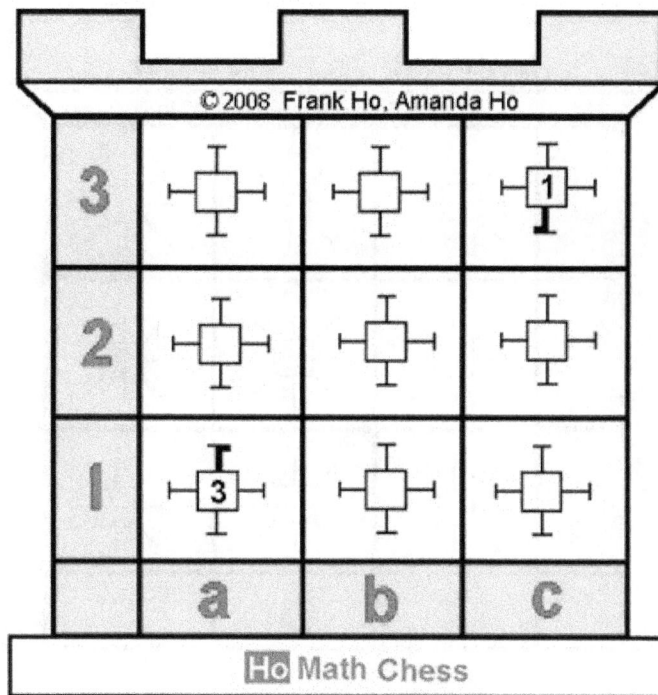

© 2008 Frank Ho, Amanda Ho

Ho Math Chess

Frankho ChessDoku™ # 5

Frankho ChessDoku™ is solved by using one or more operators of addition, subtraction, multiplication, or division after following chess moves and logic.

Rule All the digits 1 to 3 must appear exactly once in every row and column. The number appears in the bottom right-hand corner is the end result calculated according to arithmetic operator(s) and chess move(s) as indicated by darker arrow(s).

King

The following figure shows king moves by following its darker line segment(s).

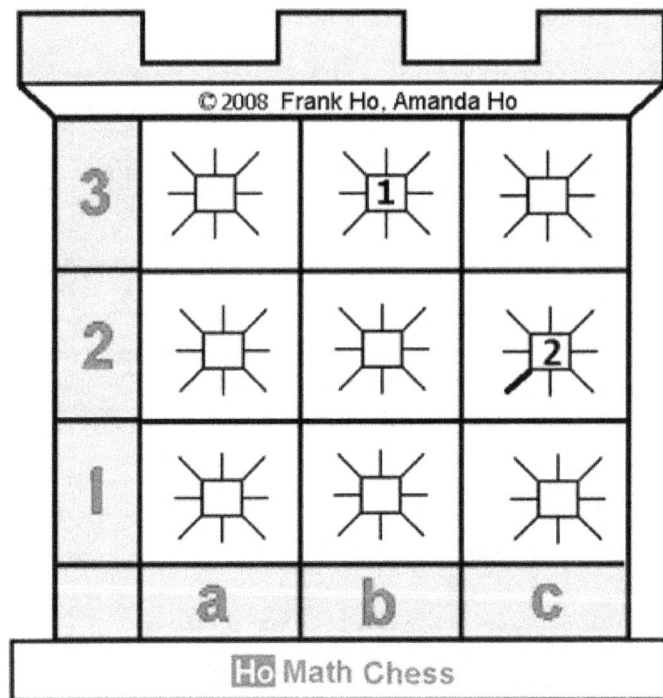

Frankho ChessDoku™ # 6

Frankho ChessDoku™ is solved by using one or more operators of addition, subtraction, multiplication, or division after following chess moves and logic.

Rule All the digits 1 to 3 must appear exactly once in every row and column. The number appears in the bottom right-hand corner is the end result calculated according to arithmetic operator(s) and chess move(s) as indicated by darker arrow(s).

Queen

The following figure shows how queen moves by following its darker line segment(s).

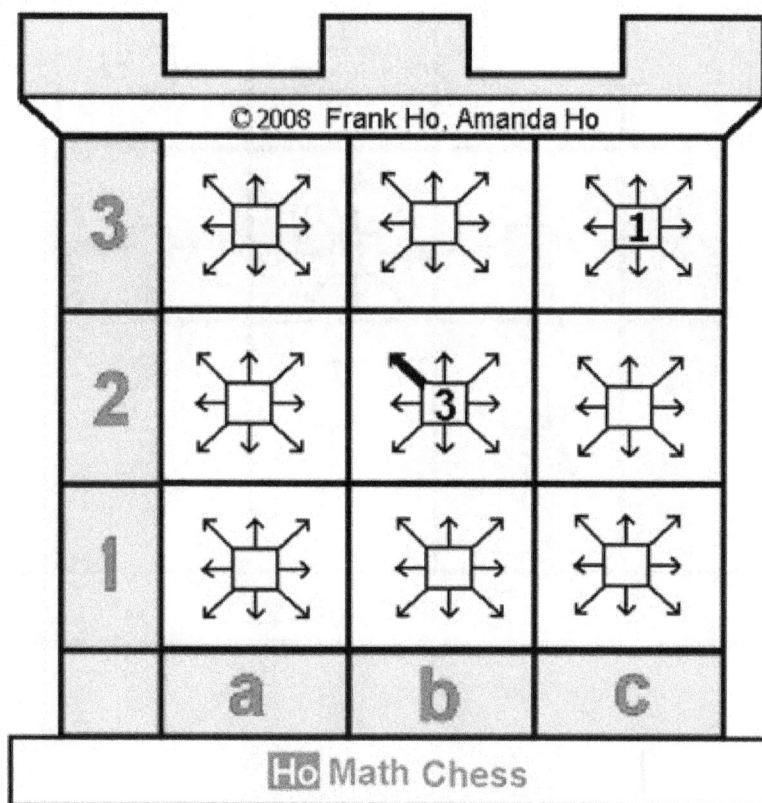

Student's name _____ Date _____

Frankho ChessDoku™ # 7

Frankho ChessDoku™ is solved by using one or more operators of addition, subtraction, multiplication, or division after following chess moves and logic.

Rule All the digits 1 to 3 must appear exactly once in every row and column. The number appears in the bottom right-hand corner is the end result calculated according to arithmetic operator(s) and chess move(s) as indicated by darker arrow(s).

© 2008 Frank Ho, Amanda Ho

Ho Math Chess

Frankho ChessDoku™ # 8

Frankho ChessDoku™ is solved by using one or more operators of addition, subtraction, multiplication, or division after following chess moves and logic.

Rule All the digits 1 to 3 must appear exactly once in every row and column. The number appears in the bottom right-hand corner is the end result calculated according to arithmetic operator(s) and chess move(s) as indicated by darker arrow(s).

Frankho ChessDoku™ # 9

Frankho ChessDoku™ is solved by using one or more operators of addition, subtraction, multiplication, or division after following chess moves and logic.

Rule All the digits 1 to 3 must appear exactly once in every row and column. The number appears in the bottom right-hand corner is the end result calculated according to arithmetic operator(s) and chess move(s) as indicated by darker arrow(s).

© 2008 Frank Ho, Amanda Ho

Ho Math Chess

Frankho ChessDoku™ # 10

Frankho ChessDoku™ is solved by using one or more operators of addition, subtraction, multiplication, or division after following chess moves and logic.

Rule All the digits 1 to 3 must appear exactly once in every row and column. The number appears in the bottom right-hand corner is the end result calculated according to arithmetic operator(s) and chess move(s) as indicated by darker arrow(s).

Student's name _____ Date _____

Frankho ChessDoku™ # 11

Frankho ChessDoku™ is solved by using addition, subtraction, multiplication, or division by following chess moves and logic.

Rule All the digits 1 to 3 must appear exactly once in every row and column. The number appears in the bottom right-hand corner is the end result calculated according to arithmetic operator(s) and chess move(s) as indicated by darker arrow(s).

© 2008 Frank Ho, Amanda Yang

Student's name _____ Date _____

Frankho ChessDoku™ # 12

Frankho ChessDoku™ is solved by using addition, subtraction, multiplication, or division by following chess moves and logic.

Rule All the digits 1 to 3 must appear exactly once in every row and column. The number appears in the bottom right-hand corner is the end result calculated according to arithmetic operator(s) and chess move(s) as indicated by darker arrow(s).

© 2008 Frank Ho, Amanda Yang

Ho Math and Chess

Student's name _____ Date _____

Frankho ChessDoku™ # 13

Frankho ChessDoku™ is solved by using addition, subtraction, multiplication, or division by following chess moves and logic.

Rule All the digits 1 to 3 must appear exactly once in every row and column. The number appears in the bottom right-hand corner is the end result calculated according to arithmetic operator(s) and chess move(s) as indicated by darker arrow(s).

© 2008 Frank Ho, Amanda Yang

Ho Math and Chess

18

Student's name _____Date _____

Frankho ChessDoku™ # 14

Frankho ChessDoku™ is solved by using addition, subtraction, multiplication, or division by following chess moves and logic.

Rule All the digits 1 to 3 must appear exactly once in every row and column. The number appears in the bottom right-hand corner is the end result calculated according to arithmetic operator(s) and chess move(s) as indicated by darker arrow(s).

Knight

© 2008 Frank Ho, Amanda Yang

Ho Math and Chess

Student's name _____ Date _____

Frankho ChessDoku™ # 15

Frankho ChessDoku™ is solved by using addition, subtraction, multiplication, or division by following chess moves and logic.

Rule All the digits 1 to 3 must appear exactly once in every row and column. The number appears in the bottom right-hand corner is the end result calculated according to arithmetic operator(s) and chess move(s) as indicated by darker arrow(s).

©2008 Frank Ho, Amanda Yang

Frankho ChessDoku™ # 16

Frankho ChessDoku™ is solved by using addition, subtraction, multiplication, or division by following chess moves and logic.

Rule All the digits 1 to 3 must appear exactly once in every row and column. The number appears in the bottom right-hand corner is the end result calculated according to arithmetic operator(s) and chess move(s) as indicated by darker arrow(s).

Student's name _____ Date _____

Frankho ChessDoku™ # 17

Frankho ChessDoku™ is solved by using addition, subtraction, multiplication, or division by following chess moves and logic.

Rule All the digits 1 to 3 must appear exactly once in every row and column. The number appears in the bottom right-hand corner is the end result calculated according to arithmetic operator(s) and chess move(s) as indicated by darker arrow(s).

© 2008 Frank Ho, Amanda Yang

Frankho ChessDoku™ # 18

Frankho ChessDoku™ is solved by using addition, subtraction, multiplication, or division by following chess moves and logic.

Rule All the digits 1 to 3 must appear exactly once in every row and column. The number appears in the bottom right-hand corner is the end result calculated according to arithmetic operator(s) and chess move(s) as indicated by darker arrow(s).

© 2008 Frank Ho, Amanda Yang

Ho Math and Chess

Student's name _____ Date _____

Frankho ChessDoku™ # 19

Frankho ChessDoku™ is solved by using addition, subtraction, multiplication, or division by following chess moves and logic.

Rule All the digits 1 to 3 must appear exactly once in every row and column. The number appears in the bottom right-hand corner is the end result calculated according to arithmetic operator(s) and chess move(s) as indicated by darker arrow(s).

© 2008 Frank Ho, Amanda Yang

Ho Math and Chess

Student's name _____ Date _____

Frankho ChessDoku™ # 20

Frankho ChessDoku™ is solved by using addition, subtraction, multiplication, or division by following chess moves and logic.

Rule All the digits 1 to 3 must appear exactly once in every row and column. The number appears in the bottom right-hand corner is the end result calculated according to arithmetic operator(s) and chess move(s) as indicated by darker arrow(s).

© 2008 Frank Ho, Amanda Yang

Ho Math and Chess

Student's name _____ Date _____

Frankho ChessDoku™ # 21

Frankho ChessDoku™ is solved by using addition, subtraction, multiplication, or division by following chess moves and logic.

Rule: All the digits 1 to 3 must appear exactly once in every row and column. The number appears in the bottom right-hand corner is the end result calculated according to arithmetic operator(s) and chess move(s) as indicated by darker arrow(s).

© 2008 Frank Ho, Amanda Yang

Frankho ChessDoku™ # 22

Frankho ChessDoku™ is solved by using addition, subtraction, multiplication, or division by following chess moves and logic.

Rule: All the digits 1 to 3 must appear exactly once in every row and column. The number appears in the bottom right-hand corner is the end result calculated according to arithmetic operator(s) and chess move(s) as indicated by darker arrow(s).

© 2008 Frank Ho, Amanda Yang

Student's name _____ Date _____

Frankho ChessDoku™ # 23

Frankho ChessDoku™ is solved by using addition, subtraction, multiplication, or division by following chess moves and logic.

Rule: All the digits 1 to 3 must appear exactly once in every row and column. The number appears in the bottom right-hand corner is the end result calculated according to arithmetic operator(s) and chess move(s) as indicated by darker arrow(s).

© 2008 Frank Ho, Amanda Yang

Ho Math and Chess

Frankho ChessDoku™ # 24

Frankho ChessDoku™ is solved by using addition, subtraction, multiplication, or division by following chess moves and logic.

Rule: All the digits 1 to 3 must appear exactly once in every row and column. The number appears in the bottom right-hand corner is the end result calculated according to arithmetic operator(s) and chess move(s) as indicated by darker arrow(s).

©2008 Frank Ho, Amanda Yang

Ho Math and Chess

Student's name _____ Date _____

Frankho ChessDoku™ # 25

Frankho ChessDoku™ is solved by using addition, subtraction, multiplication, or division by following chess moves and logic.

Rule: All the digits 1 to 3 must appear exactly once in every row and column. The number appears in the bottom right-hand corner is the end result calculated according to arithmetic operator(s) and chess move(s) as indicated by darker arrow(s).

Frankho ChessDoku™ # 26

Frankho ChessDoku™ is solved by using addition, subtraction, multiplication, or division by following chess moves and logic.

Rule: All the digits 1 to 3 must appear exactly once in every row and column. The number appears in the bottom right-hand corner is the end result calculated according to arithmetic operator(s) and chess move(s) as indicated by darker arrow(s).

© 2008 Frank Ho, Amanda Yang

Student's name _____ Date _____

Frankho ChessDoku™ # 27

Frankho ChessDoku™ is solved by using addition, subtraction, multiplication, or division by following chess moves and logic.

Rule: All the digits 1 to 3 must appear exactly once in every row and column. The number appears in the bottom right-hand corner is the end result calculated according to arithmetic operator(s) and chess move(s) as indicated by darker arrow(s).

©2008 Frank Ho, Amanda Yang

Ho Math and Chess

Student's name _____ Date _____

Frankho ChessDoku™ # 28

Frankho ChessDoku™ is solved by using addition, subtraction, multiplication, or division by following chess moves and logic.

Rule All the digits 1 to 3 must appear exactly once in every row and column. The number appears in the bottom right-hand corner is the end result calculated according to arithmetic operator(s) and chess move(s) as indicated by darker arrow(s).

© 2008 Frank Ho, Amanda Yang

Student's name _____ Date _____

Frankho ChessDoku™ # 29

Frankho ChessDoku™ is solved by using addition, subtraction, multiplication, or division by following chess moves and logic.

Rule: All the digits 1 to 3 must appear exactly once in every row and column. The number appears in the bottom right-hand corner is the end result calculated according to arithmetic operator(s) and chess move(s) as indicated by darker arrow(s).

©2008 Frank Ho, Amanda Yang

Frankho ChessDoku™ # 30

Frankho ChessDoku™ is solved by using addition, subtraction, multiplication, or division by following chess moves and logic.

Rule: All the digits 1 to 3 must appear exactly once in every row and column. The number appears in the bottom right-hand corner is the end result calculated according to arithmetic operator(s) and chess move(s) as indicated by darker arrow(s).

© 2008 Frank Ho, Amanda Yang

Student's name _____   Date _____

Frankho ChessDoku™ # 31

Frankho ChessDoku™ is solved by using addition, subtraction, multiplication, or division by following chess moves and logic.

Rule: All the digits 1 to 3 must appear exactly once in every row and column. The number appears in the bottom right-hand corner is the end result calculated according to arithmetic operator(s) and chess move(s) as indicated by darker arrow(s).

© 2008 Frank Ho, Amanda Yang

Student's name _____ Date _____

Frankho ChessDoku™ # 32

Rule: All the digits 1 to 3 must appear exactly once in every row and column. The number appears in the bottom right-hand corner is the end result calculated according to arithmetic operator(s) and chess move(s) as indicated by darker arrow(s).

© 2008 Frank Ho, Amanda Yang

Frankho ChessDoku™ # 33

Frankho ChessDoku™ is solved by using addition, subtraction, multiplication, or division by following chess moves and logic.

Rule: All the digits 1 to 3 must appear exactly once in every row and column. The number appears in the bottom right-hand corner is the end result calculated according to arithmetic operator(s) and chess move(s) as indicated by darker arrow(s).

©2008 Frank Ho, Amanda Yang

Ho Math and Chess

Frankho ChessDoku™ # 34

Frankho ChessDoku™ is solved by using addition, subtraction, multiplication, or division by following chess moves and logic.

Rule: All the digits 1 to 3 must appear exactly once in every row and column. The number appears in the bottom right-hand corner is the end result calculated according to arithmetic operator(s) and chess move(s) as indicated by darker arrow(s).

© 2008 Frank Ho, Amanda Yang

Student's name _____ Date _____

Frankho ChessDoku™ # 35

Frankho ChessDoku™ is solved by using addition, subtraction, multiplication, or division by following chess moves and logic.

Rule: All the digits 1 to 3 must appear exactly once in every row and column. The number appears in the bottom right-hand corner is the end result calculated according to arithmetic operator(s) and chess move(s) as indicated by darker arrow(s).

© 2008 Frank Ho, Amanda Yang

Frankho ChessDoku™ # 36

Frankho ChessDoku™ is solved by using addition, subtraction, multiplication, or division by following chess moves and logic.

Rule: All the digits 1 to 3 must appear exactly once in every row and column. The number appears in the bottom right-hand corner is the end result calculated according to arithmetic operator(s) and chess move(s) as indicated by darker arrow(s).

© 2008 Frank Ho, Amanda Yang

Frankho ChessDoku™ # 37

Frankho ChessDoku™ is solved by using addition, subtraction, multiplication, or division by following chess moves and logic.

Rule: All the digits 1 to 3 must appear exactly once in every row and column. The number appears in the bottom right-hand corner is the end result calculated according to arithmetic operator(s) and chess move(s) as indicated by darker arrow(s).

© 2008 Frank Ho, Amanda Yang

Ho Math and Chess

Student's name _____ Date _____

Frankho ChessDoku™ # 38

Frankho ChessDoku™ is solved by using addition, subtraction, multiplication, or division by following chess moves and logic.

Rule: All the digits 1 to 3 must appear exactly once in every row and column. The number appears in the bottom right-hand corner is the end result calculated according to arithmetic operator(s) and chess move(s) as indicated by darker arrow(s).

© 2008 Frank Ho, Amanda Yang

Frankho ChessDoku™ # 39

Frankho ChessDoku™ is solved by using addition, subtraction, multiplication, or division by following chess moves and logic.

Rule: All the digits 1 to 3 must appear exactly once in every row and column. The number appears in the bottom right-hand corner is the end result calculated according to arithmetic operator(s) and chess move(s) as indicated by darker arrow(s).

© 2008 Frank Ho, Amanda Yang

Ho Math and Chess

Frankho ChessDoku™ # 40

Frankho ChessDoku™ is solved by using addition, subtraction, multiplication, or division by following chess moves and logic.

Rule: All the digits 1 to 3 must appear exactly once in every row and column. The number appears in the bottom right-hand corner is the end result calculated according to arithmetic operator(s) and chess move(s) as indicated by darker arrow(s).

© 2008 Frank Ho, Amanda Yang

Frankho ChessDoku™ # 41

Frankho ChessDoku™ is solved by using addition, subtraction, multiplication, or division by following chess moves and logic.

Rule: All the digits 1 to 3 must appear exactly once in every row and column. The number appears in the bottom right-hand corner is the end result calculated according to arithmetic operator(s) and chess move(s) as indicated by darker arrow(s).

© 2008 Frank Ho, Amanda Yang

Frankho ChessDoku™ # 42

Frankho ChessDoku™ is solved by using addition, subtraction, multiplication, or division by following chess moves and logic.

Rule: All the digits 1 to 3 must appear exactly once in every row and column. The number appears in the bottom right-hand corner is the end result calculated according to arithmetic operator(s) and chess move(s) as indicated by darker arrow(s).

© 2008 Frank Ho, Amanda Yang

Ho Math and Chess

Frankho ChessDoku™ # 43

Frankho ChessDoku™ is solved by using addition, subtraction, multiplication, or division by following chess moves and logic.

Rule: All the digits 1 to 3 must appear exactly once in every row and column. The number appears in the bottom right-hand corner is the end result calculated according to arithmetic operator(s) and chess move(s) as indicated by darker arrow(s).

© 2008 Frank Ho, Amanda Yang

Ho Math and Chess

Student's name _____ Date _____

Frankho ChessDoku™ # 44

Frankho ChessDoku™ is solved by using addition, subtraction, multiplication, or division by following chess moves and logic.

Rule: All the digits 1 to 3 must appear exactly once in every row and column. The number appears in the bottom right-hand corner is the end result calculated according to arithmetic operator(s) and chess move(s) as indicated by darker arrow(s).

© 2008 Frank Ho, Amanda Yang

Ho Math and Chess

Frankho ChessDoku™ # 45

Rule: All the digits 1 to 3 must appear exactly once in every row and column. The number appears in the bottom right-hand corner is the end result calculated according to arithmetic operator(s) and chess move(s) as indicated by darker arrow(s).

© 2008 Frank Ho, Amanda Yang

Frankho ChessDoku™ # 46

Frankho ChessDoku™ is solved by using addition, subtraction, multiplication, or division by following chess moves and logic.

Rule: All the digits 1 to 3 must appear exactly once in every row and column. The number appears in the bottom right-hand corner is the end result calculated according to arithmetic operator(s) and chess move(s) as indicated by darker arrow(s).

© 2008 Frank Ho, Amanda Yang

Frankho ChessDoku™ # 47

Frankho ChessDoku™ is solved by using addition, subtraction, multiplication, or division by following chess moves and logic.

Rule: All the digits 1 to 3 must appear exactly once in every row and column. The number appears in the bottom right-hand corner is the end result calculated according to arithmetic operator(s) and chess move(s) as indicated by darker arrow(s).

© 2008 Frank Ho, Amanda Yang

Student's name _____ Date _____

Frankho ChessDoku™ # 48

Frankho ChessDoku™ is solved by using addition, subtraction, multiplication, or division by following chess moves and logic.

Rule: All the digits 1 to 3 must appear exactly once in every row and column. The number appears in the bottom right-hand corner is the end result calculated according to arithmetic operator(s) and chess move(s) as indicated by darker arrow(s).

© 2008 Frank Ho, Amanda Yang

Ho Math and Chess

Student's name _____ Date _____

Frankho ChessDoku™ # 49

Frankho ChessDoku™ is solved by using addition, subtraction, multiplication, or division by following chess moves and logic.

Rule: All the digits 1 to 3 must appear exactly once in every row and column. The number appears in the bottom right-hand corner is the end result calculated according to arithmetic operator(s) and chess move(s) as indicated by darker arrow(s).

© 2008 Frank Ho, Amanda Yang

Ho Math and Chess

Frankho ChessDoku™ # 50

Frankho ChessDoku™ is solved by using addition, subtraction, multiplication, or division by following chess moves and logic.

Rule: All the digits 1 to 3 must appear exactly once in every row and column. The number appears in the bottom right-hand corner is the end result calculated according to arithmetic operator(s) and chess move(s) as indicated by darker arrow(s).

© 2008 Frank Ho, Amanda Yang

Ho Math and Chess

Frankho ChessDoku™ # 51

Frankho ChessDoku™ is solved by using addition, subtraction, multiplication, or division by following chess moves and logic.

Rule: All the digits 1 to 3 must appear exactly once in every row and column. The number appears in the bottom right-hand corner is the end result calculated according to arithmetic operator(s) and chess move(s) as indicated by darker arrow(s).

© 2008 Frank Ho, Amanda Yang

Frankho ChessDoku™ # 52

Frankho ChessDoku™ is solved by using addition, subtraction, multiplication, or division by following chess moves and logic.

Rule: All the digits 1 to 3 must appear exactly once in every row and column. The number appears in the bottom right-hand corner is the end result calculated according to arithmetic operator(s) and chess move(s) as indicated by darker arrow(s).

© 2008 Frank Ho, Amanda Yang

Student's name _____   Date _____

Frankho ChessDoku™ # 53

Frankho ChessDoku™ is solved by using addition, subtraction, multiplication, or division by following chess moves and logic.

Rule: All the digits 1 to 3 must appear exactly once in every row and column. The number appears in the bottom right-hand corner is the end result calculated according to arithmetic operator(s) and chess move(s) as indicated by darker arrow(s).

© 2008 Frank Ho, Amanda Yang

Frankho ChessDoku™ # 54

Frankho ChessDoku™ is solved by using addition, subtraction, multiplication, or division by following chess moves and logic.

Rule: All the digits 1 to 3 must appear exactly once in every row and column. The number appears in the bottom right-hand corner is the end result calculated according to arithmetic operator(s) and chess move(s) as indicated by darker arrow(s).

© 2008 Frank Ho, Amanda Yang

Frankho ChessDoku™ # 55

Frankho ChessDoku™ is solved by using addition, subtraction, multiplication, or division by following chess moves and logic.

Rule: All the digits 1 to 3 must appear exactly once in every row and column. The number appears in the bottom right-hand corner is the end result calculated according to arithmetic operator(s) and chess move(s) as indicated by darker arrow(s).

© 2008 Frank Ho, Amanda Yang

Frankho ChessDoku™ # 56

Frankho ChessDoku™ is solved by using addition, subtraction, multiplication, or division by following chess moves and logic.

Rule: All the digits 1 to 3 must appear exactly once in every row and column. The number appears in the bottom right-hand corner is the end result calculated according to arithmetic operator(s) and chess move(s) as indicated by darker arrow(s).

© 2008 Frank Ho, Amanda Yang

Student's name _____ Date _____

Frankho ChessDoku™ # 57

Frankho ChessDoku™ is solved by using addition, subtraction, multiplication, or division by following chess moves and logic.

Rule: All the digits 1 to 3 must appear exactly once in every row and column. The number appears in the bottom right-hand corner is the end result calculated according to arithmetic operator(s) and chess move(s) as indicated by darker arrow(s).

© 2008 Frank Ho, Amanda Yang

Ho Math and Chess

Frankho ChessDoku™ # 58

Frankho ChessDoku™ is solved by using addition, subtraction, multiplication, or division by following chess moves and logic.

Rule: All the digits 1 to 3 must appear exactly once in every row and column. The number appears in the bottom right-hand corner is the end result calculated according to arithmetic operator(s) and chess move(s) as indicated by darker arrow(s).

© 2008 Frank Ho, Amanda Yang

Frankho ChessDoku™ # 59

Frankho ChessDoku™ is solved by using addition, subtraction, multiplication, or division by following chess moves and logic.

Rule: All the digits 1 to 3 must appear exactly once in every row and column. The number appears in the bottom right-hand corner is the end result calculated according to arithmetic operator(s) and chess move(s) as indicated by darker arrow(s).

© 2008 Frank Ho, Amanda Yang

Student's name _____ Date _____

Frankho ChessDoku™ # 60

Frankho ChessDoku™ is solved by using addition, subtraction, multiplication, or division by following chess moves and logic.

Rule: All the digits 1 to 3 must appear exactly once in every row and column. The number appears in the bottom right-hand corner is the end result calculated according to arithmetic operator(s) and chess move(s) as indicated by darker arrow(s).

© 2008 Frank Ho, Amanda Yang

Frankho ChessDoku™ # 61

Frankho ChessDoku™ is solved by using addition, subtraction, multiplication, or division by following chess moves and logic.

Rule: All the digits 1 to 3 must appear exactly once in every row and column. The number appears in the bottom right-hand corner is the end result calculated according to arithmetic operator(s) and chess move(s) as indicated by darker arrow(s).

© 2008 Frank Ho, Amanda Yang

Ho Math and Chess

Student's name _____ Date _____

Frankho ChessDoku™ # 62

Frankho ChessDoku™ is solved by using addition, subtraction, multiplication, or division by following chess moves and logic.

Rule: All the digits 1 to 3 must appear exactly once in every row and column. The number appears in the bottom right-hand corner is the end result calculated according to arithmetic operator(s) and chess move(s) as indicated by darker arrow(s).

©2008 Frank Ho, Amanda Yang

Frankho ChessDoku™ # 63

Frankho ChessDoku™ is solved by using addition, subtraction, multiplication, or division by following chess moves and logic.

Rule: All the digits 1 to 3 must appear exactly once in every row and column. The number appears in the bottom right-hand corner is the end result calculated according to arithmetic operator(s) and chess move(s) as indicated by darker arrow(s).

© 2008 Frank Ho, Amanda Yang

Student's name _____ Date _____

Frankho ChessDoku™ # 64

Frankho ChessDoku™ is solved by using addition, subtraction, multiplication, or division by following chess moves and logic.

Rule: All the digits 1 to 3 must appear exactly once in every row and column. The number appears in the bottom right-hand corner is the end result calculated according to arithmetic operator(s) and chess move(s) as indicated by darker arrow(s).

© 2008 Frank Ho, Amanda Yang

Ho Math and Chess

Student's name _____ Date _____

Frankho ChessDoku™ # 65

Frankho ChessDoku™ is solved by using addition, subtraction, multiplication, or division by following chess moves and logic.

Rule: All the digits 1 to 3 must appear exactly once in every row and column. The number appears in the bottom right-hand corner is the end result calculated according to arithmetic operator(s) and chess move(s) as indicated by darker arrow(s).

©2008 Frank Ho, Amanda Yang

Frankho ChessDoku™ # 66

Rule: All the digits 1 to 3 must appear exactly once in every row and column. The number appears in the bottom right-hand corner is the end result calculated according to arithmetic operator(s) and chess move(s) as indicated by darker arrow(s).

© 2008 Frank Ho, Amanda Yang

Frankho ChessDoku™ # 67

Frankho ChessDoku™ is solved by using addition, subtraction, multiplication, or division by following chess moves and logic.

Rule: All the digits 1 to 3 must appear exactly once in every row and column. The number appears in the bottom right-hand corner is the end result calculated according to arithmetic operator(s) and chess move(s) as indicated by darker arrow(s).

© 2008 Frank Ho, Amanda Yang

Ho Math and Chess

Frankho ChessDoku™ # 68

Frankho ChessDoku™ is solved by using addition, subtraction, multiplication, or division by following chess moves and logic.

Rule: All the digits 1 to 3 must appear exactly once in every row and column. The number appears in the bottom right-hand corner is the end result calculated according to arithmetic operator(s) and chess move(s) as indicated by darker arrow(s).

© 2008 Frank Ho, Amanda Yang

Student's name _____ Date _____

Frankho ChessDoku™ # 69

Frankho ChessDoku™ is solved by using addition, subtraction, multiplication, or division by following chess moves and logic.

Rule: All the digits 1 to 3 must appear exactly once in every row and column. The number appears in the bottom right-hand corner is the end result calculated according to arithmetic operator(s) and chess move(s) as indicated by darker arrow(s).

©2008 Frank Ho, Amanda Yang

Student's name _____ Date _____

Frankho ChessDoku™ # 70

Frankho ChessDoku™ is solved by using addition, subtraction, multiplication, or division by following chess moves and logic.

Rule: All the digits 1 to 3 must appear exactly once in every row and column. The number appears in the bottom right-hand corner is the end result calculated according to arithmetic operator(s) and chess move(s) as indicated by darker arrow(s).

© 2008 Frank Ho, Amanda Yang

Frankho ChessDoku™ # 71

Frankho ChessDoku™ is solved by using addition, subtraction, multiplication, or division by following chess moves and logic.

Rule: All the digits 1 to 3 must appear exactly once in every row and column. The number appears in the bottom right-hand corner is the end result calculated according to arithmetic operator(s) and chess move(s) as indicated by darker arrow(s).

© 2008 Frank Ho, Amanda Yang

Student's name _____ Date _____

Frankho ChessDoku™ # 72

Frankho ChessDoku™ is solved by using addition, subtraction, multiplication, or division by following chess moves and logic.

Rule: All the digits 1 to 3 must appear exactly once in every row and column. The number appears in the bottom right-hand corner is the end result calculated according to arithmetic operator(s) and chess move(s) as indicated by darker arrow(s).

© 2008 Frank Ho, Amanda Yang

Ho Math and Chess

Frankho ChessDoku™ # 73

Frankho ChessDoku™ is solved by using addition, subtraction, multiplication, or division by following chess moves and logic.

Rule: All the digits 1 to 3 must appear exactly once in every row and column. The number appears in the bottom right-hand corner is the end result calculated according to arithmetic operator(s) and chess move(s) as indicated by darker arrow(s).

Frankho ChessDoku™ # 74

Frankho ChessDoku™ is solved by using addition, subtraction, multiplication, or division by following chess moves and logic.

Rule: All the digits 1 to 3 must appear exactly once in every row and column. The number appears in the bottom right-hand corner is the end result calculated according to arithmetic operator(s) and chess move(s) as indicated by darker arrow(s).

©2008 Frank Ho, Amanda Yang

Frankho ChessDoku™ # 75

Frankho ChessDoku™ is solved by using addition, subtraction, multiplication, or division by following chess moves and logic.

Rule: All the digits 1 to 3 must appear exactly once in every row and column. The number appears in the bottom right-hand corner is the end result calculated according to arithmetic operator(s) and chess move(s) as indicated by darker arrow(s).

© 2008 Frank Ho, Amanda Yang

Ho Math and Chess

Frankho ChessDoku™ # 76

Frankho ChessDoku™ is solved by using addition, subtraction, multiplication, or division by following chess moves and logic.

Rule: All the digits 1 to 3 must appear exactly once in every row and column. The number appears in the bottom right-hand corner is the end result calculated according to arithmetic operator(s) and chess move(s) as indicated by darker arrow(s).

© 2008 Frank Ho, Amanda Yang

Ho Math and Chess

Frankho ChessDoku™ # 77

Frankho ChessDoku™ is solved by using addition, subtraction, multiplication, or division by following chess moves and logic.

Rule: All the digits 1 to 3 must appear exactly once in every row and column. The number appears in the bottom right-hand corner is the end result calculated according to arithmetic operator(s) and chess move(s) as indicated by darker arrow(s).

© 2008 Frank Ho, Amanda Yang

Frankho ChessDoku™ # 78

Frankho ChessDoku™ is solved by using addition, subtraction, multiplication, or division by following chess moves and logic.

Rule: All the digits 1 to 3 must appear exactly once in every row and column. The number appears in the bottom right-hand corner is the end result calculated according to arithmetic operator(s) and chess move(s) as indicated by darker arrow(s).

© 2008 Frank Ho, Amanda Yang

Ho Math and Chess

Frankho ChessDoku™ # 79

Frankho ChessDoku™ is solved by using addition, subtraction, multiplication, or division by following chess moves and logic.

Rule: All the digits 1 to 3 must appear exactly once in every row and column. The number appears in the bottom right-hand corner is the end result calculated according to arithmetic operator(s) and chess move(s) as indicated by darker arrow(s).

© 2008 Frank Ho, Amanda Yang

Ho Math and Chess

Student's name _____ Date _____

Frankho ChessDoku™ # 80

Frankho ChessDoku™ is solved by using addition, subtraction, multiplication, or division by following chess moves and logic.

Rule: All the digits 1 to 3 must appear exactly once in every row and column. The number appears in the bottom right-hand corner is the end result calculated according to arithmetic operator(s) and chess move(s) as indicated by darker arrow(s).

© 2008 Frank Ho, Amanda Yang

Ho Math and Chess

Student's name _____   Date _____

Frankho ChessDoku™ # 81

Frankho ChessDoku™ is solved by using addition, subtraction, multiplication, or division by following chess moves and logic.

Rule: All the digits 1 to 3 must appear exactly once in every row and column. The number appears in the bottom right-hand corner is the end result calculated according to arithmetic operator(s) and chess move(s) as indicated by darker arrow(s).

© 2008 Frank Ho, Amanda Yang

Frankho ChessDoku™ # 82

Frankho ChessDoku™ is solved by using addition, subtraction, multiplication, or division by following chess moves and logic.

Rule: All the digits 1 to 3 must appear exactly once in every row and column. The number appears in the bottom right-hand corner is the end result calculated according to arithmetic operator(s) and chess move(s) as indicated by darker arrow(s).

©2008 Frank Ho, Amanda Yang

Frankho ChessDoku™ # 83

Frankho ChessDoku™ is solved by using addition, subtraction, multiplication, or division by following chess moves and logic.

Rule: All the digits 1 to 3 must appear exactly once in every row and column. The number appears in the bottom right-hand corner is the end result calculated according to arithmetic operator(s) and chess move(s) as indicated by darker arrow(s).

© 2008 Frank Ho, Amanda Yang

Ho Math and Chess

Frankho ChessDoku™ # 84

Frankho ChessDoku™ is solved by using addition, subtraction, multiplication, or division by following chess moves and logic.

Rule: All the digits 1 to 3 must appear exactly once in every row and column. The number appears in the bottom right-hand corner is the end result calculated according to arithmetic operator(s) and chess move(s) as indicated by darker arrow(s).

© 2008 Frank Ho, Amanda Yang

Ho Math and Chess

Frankho ChessDoku™ # 85

Frankho ChessDoku™ is solved by using addition, subtraction, multiplication, or division by following chess moves and logic.

Rule: All the digits 1 to 3 must appear exactly once in every row and column. The number appears in the bottom right-hand corner is the end result calculated according to arithmetic operator(s) and chess move(s) as indicated by darker arrow(s).

© 2008 Frank Ho, Amanda Yang

Frankho ChessDoku™ # 86

Frankho ChessDoku™ is solved by using addition, subtraction, multiplication, or division by following chess moves and logic.

Rule: All the digits 1 to 3 must appear exactly once in every row and column. The number appears in the bottom right-hand corner is the end result calculated according to arithmetic operator(s) and chess move(s) as indicated by darker arrow(s).

Frankho ChessDoku™ # 87

Frankho ChessDoku™ is solved by using addition, subtraction, multiplication, or division by following chess moves and logic.

Rule: All the digits 1 to 3 must appear exactly once in every row and column. The number appears in the bottom right-hand corner is the end result calculated according to arithmetic operator(s) and chess move(s) as indicated by darker arrow(s).

© 2008 Frank Ho, Amanda Yang

Ho Math and Chess

Student's name _____ Date _____

Frankho ChessDoku™ # 88

Frankho ChessDoku™ is solved by using addition, subtraction, multiplication, or division by following chess moves and logic.

Rule: All the digits 1 to 3 must appear exactly once in every row and column. The number appears in the bottom right-hand corner is the end result calculated according to arithmetic operator(s) and chess move(s) as indicated by darker arrow(s).

© 2008 Frank Ho, Amanda Yang

Ho Math and Chess

Frankho ChessDoku™ # 89

Frankho ChessDoku™ is solved by using addition, subtraction, multiplication, or division by following chess moves and logic.

Rule: All the digits 1 to 3 must appear exactly once in every row and column. The number appears in the bottom right-hand corner is the end result calculated according to arithmetic operator(s) and chess move(s) as indicated by darker arrow(s).

© 2008 Frank Ho, Amanda Yang

Ho Math and Chess

Student's name _____ Date _____

Frankho ChessDoku™ # 90

Frankho ChessDoku™ is solved by using addition, subtraction, multiplication, or division by following chess moves and logic.

Rule: All the digits 1 to 3 must appear exactly once in every row and column. The number appears in the bottom right-hand corner is the end result calculated according to arithmetic operator(s) and chess move(s) as indicated by darker arrow(s).

© 2008 Frank Ho, Amanda Yang

Ho Math and Chess

Student's name _____ Date _____

Frankho ChessDoku™ # 91

RULES:

All the digits 1 to 3 must appear in every row and column. The number appears in the bottom right-hand corner is the end result calculated according to operator(s) and chess move(s).

© 2008 Frank Ho, Amanda Yang

Frankho ChessDoku™ # 92

Frankho ChessDoku™ is solved by using addition, subtraction, multiplication, or division by following chess moves and logic.

Rule: All the digits 1 to 3 must appear exactly once in every row and column. The number appears in the bottom right-hand corner is the end result calculated according to arithmetic operator(s) and chess move(s) as indicated by darker arrow(s).

© 2008 Frank Ho, Amanda Yang

Frankho ChessDoku™ # 93

Frankho ChessDoku™ is solved by using addition, subtraction, multiplication, or division by following chess moves and logic.

Rule: All the digits 1 to 3 must appear exactly once in every row and column. The number appears in the bottom right-hand corner is the end result calculated according to arithmetic operator(s) and chess move(s) as indicated by darker arrow(s).

© 2008 Frank Ho, Amanda Yang

Ho Math and Chess

Student's name _____ Date _____

Frankho ChessDoku™ # 94

Frankho ChessDoku™ is solved by using addition, subtraction, multiplication, or division by following chess moves and logic.

Rule: All the digits 1 to 3 must appear exactly once in every row and column. The number appears in the bottom right-hand corner is the end result calculated according to arithmetic operator(s) and chess move(s) as indicated by darker arrow(s).

© 2008 Frank Ho, Amanda Yang

Ho Math and Chess

Frankho ChessDoku™ # 95

Frankho ChessDoku™ is solved by using addition, subtraction, multiplication, or division by following chess moves and logic.

Rule: All the digits 1 to 3 must appear exactly once in every row and column. The number appears in the bottom right-hand corner is the end result calculated according to arithmetic operator(s) and chess move(s) as indicated by darker arrow(s).

© 2008 Frank Ho, Amanda Yang

Frankho ChessDoku™ # 96

Frankho ChessDoku™ is solved by using addition, subtraction, multiplication, or division by following chess moves and logic.

Rule: All the digits 1 to 3 must appear exactly once in every row and column. The number appears in the bottom right-hand corner is the end result calculated according to arithmetic operator(s) and chess move(s) as indicated by darker arrow(s).

© 2008 Frank Ho, Amanda Yang

Frankho ChessDoku™ # 97

Frankho ChessDoku™ is solved by using addition, subtraction, multiplication, or division by following chess moves and logic.

Rule: All the digits 1 to 3 must appear exactly once in every row and column. The number appears in the bottom right-hand corner is the end result calculated according to arithmetic operator(s) and chess move(s) as indicated by darker arrow(s).

© 2008 Frank Ho, Amanda Yang

Frankho ChessDoku™ # 98

Frankho ChessDoku™ is solved by using addition, subtraction, multiplication, or division by following chess moves and logic.

Rule: All the digits 1 to 3 must appear exactly once in every row and column. The number appears in the bottom right-hand corner is the end result calculated according to arithmetic operator(s) and chess move(s) as indicated by darker arrow(s).

© 2008 Frank Ho, Amanda Yang

Ho Math and Chess

Frankho ChessDoku™ # 99

Frankho ChessDoku™ is solved by using addition, subtraction, multiplication, or division by following chess moves and logic.

Rule: All the digits 1 to 3 must appear exactly once in every row and column. The number appears in the bottom right-hand corner is the end result calculated according to arithmetic operator(s) and chess move(s) as indicated by darker arrow(s).

© 2008 Frank Ho, Amanda Yang

Frankho ChessDoku™ # 100

Frankho ChessDoku™ is solved by using addition, subtraction, multiplication, or division by following chess moves and logic.

Rule: All the digits 1 to 3 must appear exactly once in every row and column. The number appears in the bottom right-hand corner is the end result calculated according to arithmetic operator(s) and chess move(s) as indicated by darker arrow(s).

© 2008 Frank Ho, Amanda Yang

Ho Math and Chess

Frankho ChessDoku™ # 101

Frankho ChessDoku™ is solved by using addition, subtraction, multiplication, or division by following chess moves and logic.

Rule: All the digits 1 to 3 must appear exactly once in every row and column. The number appears in the bottom right-hand corner is the end result calculated according to arithmetic operator(s) and chess move(s) as indicated by darker arrow(s).

©2008 Frank Ho, Amanda Yang

Frankho ChessDoku™ # 102

Frankho ChessDoku™ is solved by using addition, subtraction, multiplication, or division by following chess moves and logic.

Rule: All the digits 1 to 3 must appear exactly once in every row and column. The number appears in the bottom right-hand corner is the end result calculated according to arithmetic operator(s) and chess move(s) as indicated by darker arrow(s).

© 2008 Frank Ho, Amanda Yang

Ho Math and Chess

Frankho ChessDoku™ # 103

Frankho ChessDoku™ is solved by using addition, subtraction, multiplication, or division by following chess moves and logic.

Rule: All the digits 1 to 3 must appear exactly once in every row and column. The number appears in the bottom right-hand corner is the end result calculated according to arithmetic operator(s) and chess move(s) as indicated by darker arrow(s).

©2008 Frank Ho, Amanda Yang

Frankho ChessDoku™ # 104

Frankho ChessDoku™ is solved by using addition, subtraction, multiplication, or division by following chess moves and logic.

Rule: All the digits 1 to 3 must appear exactly once in every row and column. The number appears in the bottom right-hand corner is the end result calculated according to arithmetic operator(s) and chess move(s) as indicated by darker arrow(s).

© 2008 Frank Ho, Amanda Yang

Student's name _____ Date _____

Frankho ChessDoku™ # 105

Frankho ChessDoku™ is solved by using addition, subtraction, multiplication, or division by following chess moves and logic.

Rule: All the digits 1 to 3 must appear exactly once in every row and column. The number appears in the bottom right-hand corner is the end result calculated according to arithmetic operator(s) and chess move(s) as indicated by darker arrow(s).

© 2008 Frank Ho, Amanda Yang

Frankho ChessDoku™ # 106

Frankho ChessDoku™ is solved by using addition, subtraction, multiplication, or division by following chess moves and logic.

Rule: All the digits 1 to 3 must appear exactly once in every row and column. The number appears in the bottom right-hand corner is the end result calculated according to arithmetic operator(s) and chess move(s) as indicated by darker arrow(s).

© 2008 Frank Ho, Amanda Yang

Frankho ChessDoku™ # 107

Frankho ChessDoku™ is solved by using addition, subtraction, multiplication, or division by following chess moves and logic.

Rule: All the digits 1 to 3 must appear exactly once in every row and column. The number appears in the bottom right-hand corner is the end result calculated according to arithmetic operator(s) and chess move(s) as indicated by darker arrow(s).

© 2008 Frank Ho, Amanda Yang

Frankho ChessDoku™ # 108

Frankho ChessDoku™ is solved by using addition, subtraction, multiplication, or division by following chess moves and logic.

Rule: All the digits 1 to 3 must appear exactly once in every row and column. The number appears in the bottom right-hand corner is the end result calculated according to arithmetic operator(s) and chess move(s) as indicated by darker arrow(s).

© 2008 Frank Ho, Amanda Yang

Ho Math and Chess

Frankho ChessDoku™ # 109

Frankho ChessDoku™ is solved by using addition, subtraction, multiplication, or division by following chess moves and logic.

Rule: All the digits 1 to 3 must appear exactly once in every row and column. The number appears in the bottom right-hand corner is the end result calculated according to arithmetic operator(s) and chess move(s) as indicated by darker arrow(s).

© 2008 Frank Ho, Amanda Yang

Ho Math and Chess

Frankho ChessDoku™ # 110

Frankho ChessDoku™ is solved by using addition, subtraction, multiplication, or division by following chess moves and logic.

Rule: All the digits 1 to 3 must appear exactly once in every row and column. The number appears in the bottom right-hand corner is the end result calculated according to arithmetic operator(s) and chess move(s) as indicated by darker arrow(s).

© 2008 Frank Ho, Amanda Yang

Frankho ChessDoku™ # 111

Frankho ChessDoku™ is solved by using addition, subtraction, multiplication, or division by following chess moves and logic.

Rule: All the digits 1 to 3 must appear exactly once in every row and column. The number appears in the bottom right-hand corner is the end result calculated according to arithmetic operator(s) and chess move(s) as indicated by darker arrow(s).

© 2008 Frank Ho, Amanda Yang

Student's name _____ Date _____

Frankho ChessDoku™ # 112

Frankho ChessDoku™ is solved by using addition, subtraction, multiplication, or division by following chess moves and logic.

Rule: All the digits 1 to 3 must appear exactly once in every row and column. The number appears in the bottom right-hand corner is the end result calculated according to arithmetic operator(s) and chess move(s) as indicated by darker arrow(s).

© 2008 Frank Ho, Amanda Yang

Ho Math and Chess

Frankho ChessDoku™ # 113

Frankho ChessDoku™ is solved by using addition, subtraction, multiplication, or division by following chess moves and logic.

Rule: All the digits 1 to 3 must appear exactly once in every row and column. The number appears in the bottom right-hand corner is the end result calculated according to arithmetic operator(s) and chess move(s) as indicated by darker arrow(s).

© 2008 Frank Ho, Amanda Yang

Ho Math and Chess

Frankho ChessDoku™ # 114

Frankho ChessDoku™ is solved by using addition, subtraction, multiplication, or division by following chess moves and logic.

Rule: All the digits 1 to 3 must appear exactly once in every row and column. The number appears in the bottom right-hand corner is the end result calculated according to arithmetic operator(s) and chess move(s) as indicated by darker arrow(s).

Student's name _____ Date _____

Frankho ChessDoku™ # 115

Frankho ChessDoku™ is solved by using addition, subtraction, multiplication, or division by following chess moves and logic.

Rule: All the digits 1 to 3 must appear exactly once in every row and column. The number appears in the bottom right-hand corner is the end result calculated according to arithmetic operator(s) and chess move(s) as indicated by darker arrow(s).

©2008 Frank Ho, Amanda Yang

Student's name _____ Date _____

Frankho ChessDoku™ # 116

Frankho ChessDoku™ is solved by using addition, subtraction, multiplication, or division by following chess moves and logic.

Rule: All the digits 1 to 3 must appear exactly once in every row and column. The number appears in the bottom right-hand corner is the end result calculated according to arithmetic operator(s) and chess move(s) as indicated by darker arrow(s).

© 2008 Frank Ho, Amanda Yang

Ho Math and Chess

Student's name _____ Date _____

Frankho ChessDoku™ # 117

Frankho ChessDoku™ is solved by using addition, subtraction, multiplication, or division by following chess moves and logic.

Rule: All the digits 1 to 3 must appear exactly once in every row and column. The number appears in the bottom right-hand corner is the end result calculated according to arithmetic operator(s) and chess move(s) as indicated by darker arrow(s).

© 2008 Frank Ho, Amanda Yang

Student's name _____ Date _____

Frankho ChessDoku™ # 118

RULES:

All the digits 1 to 3 must appear in every row and column. The number appears in the bottom right-hand corner is the end result calculated according to operator(s) and chess move(s).

©2008 Frank Ho, Amanda Yang

Frankho ChessDoku™ # 119

Frankho ChessDoku™ is solved by using addition, subtraction, multiplication, or division by following chess moves and logic.

Rule: All the digits 1 to 3 must appear exactly once in every row and column. The number appears in the bottom right-hand corner is the end result calculated according to arithmetic operator(s) and chess move(s) as indicated by darker arrow(s).

© 2008 Frank Ho, Amanda Yang

Frankho ChessDoku™ # 120

Frankho ChessDoku™ is solved by using addition, subtraction, multiplication, or division by following chess moves and logic.

Rule: All the digits 1 to 3 must appear exactly once in every row and column. The number appears in the bottom right-hand corner is the end result calculated according to arithmetic operator(s) and chess move(s) as indicated by darker arrow(s).

© 2008 Frank Ho, Amanda Yang

Ho Math and Chess

Student's name _____ Date _____

Frankho ChessDoku™ # 121

Frankho ChessDoku™ is solved by using addition, subtraction, multiplication, or division by following chess moves and logic.

Rule: All the digits 1 to 3 must appear exactly once in every row and column. The number appears in the bottom right-hand corner is the end result calculated according to arithmetic operator(s) and chess move(s) as indicated by darker arrow(s).

© 2008 Frank Ho, Amanda Yang

Student's name _____ Date _____

Frankho ChessDoku™ # 122

Frankho ChessDoku™ is solved by using addition, subtraction, multiplication, or division by following chess moves and logic.

Rule: All the digits 1 to 3 must appear exactly once in every row and column. The number appears in the bottom right-hand corner is the end result calculated according to arithmetic operator(s) and chess move(s) as indicated by darker arrow(s).

© 2008 Frank Ho, Amanda Yang

Frankho ChessDoku™ # 123

Frankho ChessDoku™ is solved by using addition, subtraction, multiplication, or division by following chess moves and logic.

Rule: All the digits 1 to 3 must appear exactly once in every row and column. The number appears in the bottom right-hand corner is the end result calculated according to arithmetic operator(s) and chess move(s) as indicated by darker arrow(s).

© 2008 Frank Ho, Amanda Yang

Ho Math and Chess

Frankho ChessDoku™ # 124

Frankho ChessDoku™ is solved by using addition, subtraction, multiplication, or division by following chess moves and logic.

Rule: All the digits 1 to 3 must appear exactly once in every row and column. The number appears in the bottom right-hand corner is the end result calculated according to arithmetic operator(s) and chess move(s) as indicated by darker arrow(s).

© 2008 Frank Ho, Amanda Yang

Ho Math and Chess

Frankho ChessDoku™ # 125

RULES:

All the digits 1 to 3 must appear in every row and column. The number appears in the bottom right-hand corner is the end result calculated according to operator(s) and chess move(s).

© 2008 Frank Ho, Amanda Yang

Ho Math and Chess

Frankho ChessDoku™ # 126

Frankho ChessDoku™ is solved by using addition, subtraction, multiplication, or division by following chess moves and logic.

Rule: All the digits 1 to 3 must appear exactly once in every row and column. The number appears in the bottom right-hand corner is the end result calculated according to arithmetic operator(s) and chess move(s) as indicated by darker arrow(s).

© 2008 Frank Ho, Amanda Yang

Ho Math and Chess

Frankho ChessDoku™ # 127

Frankho ChessDoku™ is solved by using addition, subtraction, multiplication, or division by following chess moves and logic.

Rule: All the digits 1 to 3 must appear exactly once in every row and column. The number appears in the bottom right-hand corner is the end result calculated according to arithmetic operator(s) and chess move(s) as indicated by darker arrow(s).

Student's name _____ Date _____

Frankho ChessDoku™ # 128

Frankho ChessDoku™ is solved by using addition, subtraction, multiplication, or division by following chess moves and logic.

Rule: All the digits 1 to 3 must appear exactly once in every row and column. The number appears in the bottom right-hand corner is the end result calculated according to arithmetic operator(s) and chess move(s) as indicated by darker arrow(s).

© 2008 Frank Ho, Amanda Yang

Ho Math and Chess

Frankho ChessDoku™ # 129

Frankho ChessDoku™ is solved by using addition, subtraction, multiplication, or division by following chess moves and logic.

Rule: All the digits 1 to 3 must appear exactly once in every row and column. The number appears in the bottom right-hand corner is the end result calculated according to arithmetic operator(s) and chess move(s) as indicated by darker arrow(s).

© 2008 Frank Ho, Amanda Yang

Ho Math and Chess

Frankho ChessDoku™ # 130

RULES:

All the digits 1 to 3 must appear in every row and column. The number appears in the bottom right-hand corner is the end result calculated according to operator(s) and chess move(s).

© 2008 Frank Ho, Amanda Yang

Ho Math and Chess

Frankho ChessDoku™ # 131

Frankho ChessDoku™ is solved by using addition, subtraction, multiplication, or division by following chess moves and logic.

Rule: All the digits 1 to 3 must appear exactly once in every row and column. The number appears in the bottom right-hand corner is the end result calculated according to arithmetic operator(s) and chess move(s) as indicated by darker arrow(s).

© 2008 Frank Ho, Amanda Yang

Student's name _____ Date _____

Frankho ChessDoku™ # 132

Frankho ChessDoku™ is solved by using addition, subtraction, multiplication, or division by following chess moves and logic.

Rule: All the digits 1 to 3 must appear exactly once in every row and column. The number appears in the bottom right-hand corner is the end result calculated according to arithmetic operator(s) and chess move(s) as indicated by darker arrow(s).

© 2008 Frank Ho, Amanda Yang

Frankho ChessDoku™ # 133

Frankho ChessDoku™ is solved by using addition, subtraction, multiplication, or division by following chess moves and logic.

Rule: All the digits 1 to 3 must appear exactly once in every row and column. The number appears in the bottom right-hand corner is the end result calculated according to arithmetic operator(s) and chess move(s) as indicated by darker arrow(s).

© 2008 Frank Ho, Amanda Yang

Frankho ChessDoku™ # 134

Frankho ChessDoku™ is solved by using addition, subtraction, multiplication, or division by following chess moves and logic.

Rule: All the digits 1 to 3 must appear exactly once in every row and column. The number appears in the bottom right-hand corner is the end result calculated according to arithmetic operator(s) and chess move(s) as indicated by darker arrow(s).

© 2008 Frank Ho, Amanda Yang

Ho Math and Chess

Frankho ChessDoku™ # 135

Frankho ChessDoku™ is solved by using addition, subtraction, multiplication, or division by following chess moves and logic.

Rule: All the digits 1 to 3 must appear exactly once in every row and column. The number appears in the bottom right-hand corner is the end result calculated according to arithmetic operator(s) and chess move(s) as indicated by darker arrow(s).

© 2008 Frank Ho, Amanda Yang

Ho Math and Chess

Frankho ChessDoku™ # 136

Rule: All the digits 1 to 3 must appear exactly once in every row and column. The number appears in the bottom right-hand corner is the end result calculated according to arithmetic operator(s) and chess move(s) as indicated by darker arrow(s).

© 2008 Frank Ho, Amanda Yang

Frankho ChessDoku™ # 137

Rule: All the digits 1 to 3 must appear exactly once in every row and column. The number appears in the bottom right-hand corner is the end result calculated according to arithmetic operator(s) and chess move(s) as indicated by darker arrow(s).

©2008 Frank Ho, Amanda Yang

Ho Math and Chess

Frankho ChessDoku™ # 138

Frankho ChessDoku™ is solved by using addition, subtraction, multiplication, or division by following chess moves and logic.

Rule: All the digits 1 to 3 must appear exactly once in every row and column. The number appears in the bottom right-hand corner is the end result calculated according to arithmetic operator(s) and chess move(s) as indicated by darker arrow(s).

© 2008 Frank Ho, Amanda Yang

Frankho ChessDoku™ # 139

Frankho ChessDoku™ is solved by using addition, subtraction, multiplication, or division by following chess moves and logic.

Rule: All the digits 1 to 3 must appear exactly once in every row and column. The number appears in the bottom right-hand corner is the end result calculated according to arithmetic operator(s) and chess move(s) as indicated by darker arrow(s).

© 2008 Frank Ho, Amanda Yang

Ho Math and Chess

Frankho ChessDoku™ # 140

Frankho ChessDoku™ is solved by using addition, subtraction, multiplication, or division by following chess moves and logic.

Rule: All the digits 1 to 3 must appear exactly once in every row and column. The number appears in the bottom right-hand corner is the end result calculated according to arithmetic operator(s) and chess move(s) as indicated by darker arrow(s).

© 2008 Frank Ho, Amanda Yang

Ho Math and Chess

Student's name _____ Date _____

Frankho ChessDoku™ # 141

Frankho ChessDoku™ is solved by using addition, subtraction, multiplication, or division by following chess moves and logic.

Rule: All the digits 1 to 3 must appear exactly once in every row and column. The number appears in the bottom right-hand corner is the end result calculated according to arithmetic operator(s) and chess move(s) as indicated by darker arrow(s).

© 2008 Frank Ho, Amanda Yang

Ho Math and Chess

Student's name _____ Date _____

Frankho ChessDoku™ # 142

Frankho ChessDoku™ is solved by using addition, subtraction, multiplication, or division by following chess moves and logic.

Rule: All the digits 1 to 3 must appear exactly once in every row and column. The number appears in the bottom right-hand corner is the end result calculated according to arithmetic operator(s) and chess move(s) as indicated by darker arrow(s).

© 2008 Frank Ho, Amanda Yang

Ho Math and Chess

Frankho ChessDoku™ # 143

RULES:

All the digits 1 to 3 must appear in every row and column. The number appears in the bottom right-hand corner is the end result calculated according to operator(s) and chess move(s).

© 2008 Frank Ho, Amanda Yang

Frankho ChessDoku™ # 144

Frankho ChessDoku™ is solved by using addition, subtraction, multiplication, or division by following chess moves and logic.

Rule: All the digits 1 to 3 must appear exactly once in every row and column. The number appears in the bottom right-hand corner is the end result calculated according to arithmetic operator(s) and chess move(s) as indicated by darker arrow(s).

Frankho ChessDoku™ # 145

Frankho ChessDoku™ is solved by using addition, subtraction, multiplication, or division by following chess moves and logic.

Rule: All the digits 1 to 3 must appear exactly once in every row and column. The number appears in the bottom right-hand corner is the end result calculated according to arithmetic operator(s) and chess move(s) as indicated by darker arrow(s).

©2008 Frank Ho, Amanda Yang

Student's name _____ Date _____

Frankho ChessDoku™ # 146

Frankho ChessDoku™ is solved by using addition, subtraction, multiplication, or division by following chess moves and logic.

Rule: All the digits 1 to 3 must appear exactly once in every row and column. The number appears in the bottom right-hand corner is the end result calculated according to arithmetic operator(s) and chess move(s) as indicated by darker arrow(s).

Frankho ChessDoku™ # 147

Frankho ChessDoku™ is solved by using addition, subtraction, multiplication, or division by following chess moves and logic.

Rule: All the digits 1 to 3 must appear exactly once in every row and column. The number appears in the bottom right-hand corner is the end result calculated according to arithmetic operator(s) and chess move(s) as indicated by darker arrow(s).

Frankho ChessDoku™ # 148

Frankho ChessDoku™ is solved by using addition, subtraction, multiplication, or division by following chess moves and logic.

Rule: All the digits 1 to 3 must appear exactly once in every row and column. The number appears in the bottom right-hand corner is the end result calculated according to arithmetic operator(s) and chess move(s) as indicated by darker arrow(s).

© 2008 Frank Ho, Amanda Yang

Frankho ChessDoku™ # 149

Frankho ChessDoku™ is solved by using addition, subtraction, multiplication, or division by following chess moves and logic.

Rule: All the digits 1 to 3 must appear exactly once in every row and column. The number appears in the bottom right-hand corner is the end result calculated according to arithmetic operator(s) and chess move(s) as indicated by darker arrow(s).

© 2008 Frank Ho, Amanda Yang

Frankho ChessDoku™ # 150

RULES:

Rule: All the digits 1 to 3 must appear exactly once in every row and column. The number appears in the bottom right-hand corner is the end result calculated according to arithmetic operator(s) and chess move(s) as indicated by darker arrow(s).

© 2008 Frank Ho, Amanda Yang

Student's name _____ Date _____

Frankho ChessDoku™ # 151

RULES:

All the digits 1 to 3 must appear in every row and column. The number appears in the bottom right-hand corner is the end result calculated according to operator(s) and chess move(s).

Student's name _____ Date _____

Frankho ChessDoku™ # 152

RULES:

All the digits 1 to 3 must appear in every row and column. The number appears in the bottom right-hand corner is the end result calculated according to operator(s) and chess move(s).

© 2008 Frank Ho, Amanda Yang

***** Part 2 – 3 Dimensional Frankho ChessDoku *****

3 Dimensional Frankho ChessDoku™ # 1

Rule: All the digits 1 to 3 must appear in every row and column but cannot repeat on the same row of the same column of each layer. The number appears in the bottom right-hand corner is the end result calculated according to operator(s) and chess move(s).

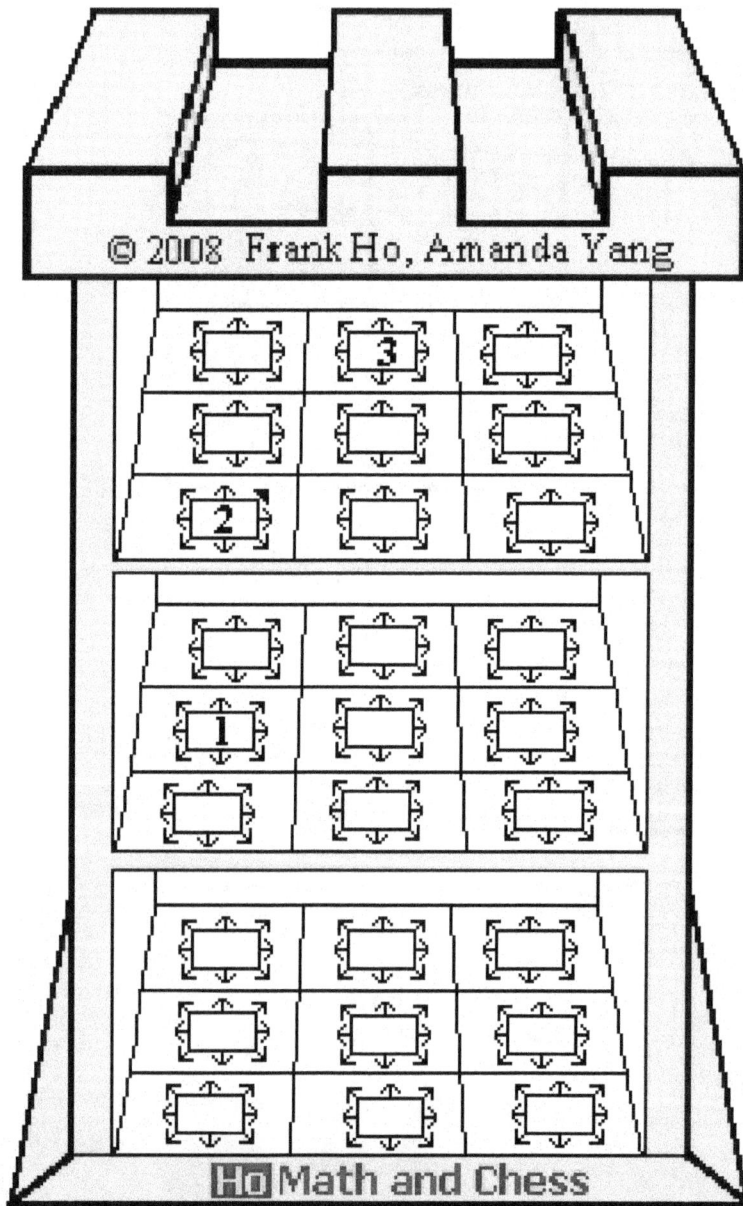

© 2008 Frank Ho, Amanda Yang

Ho Math and Chess

Student's name _____ Date _____

3 Dimensional Frankho ChessDoku™ # 2

Rule: All the digits 1 to 3 must appear in every row and column but cannot repeat on the same row of the same column of each layer. The number appears in the bottom right-hand corner is the end result calculated according to operator(s) and chess move(s).

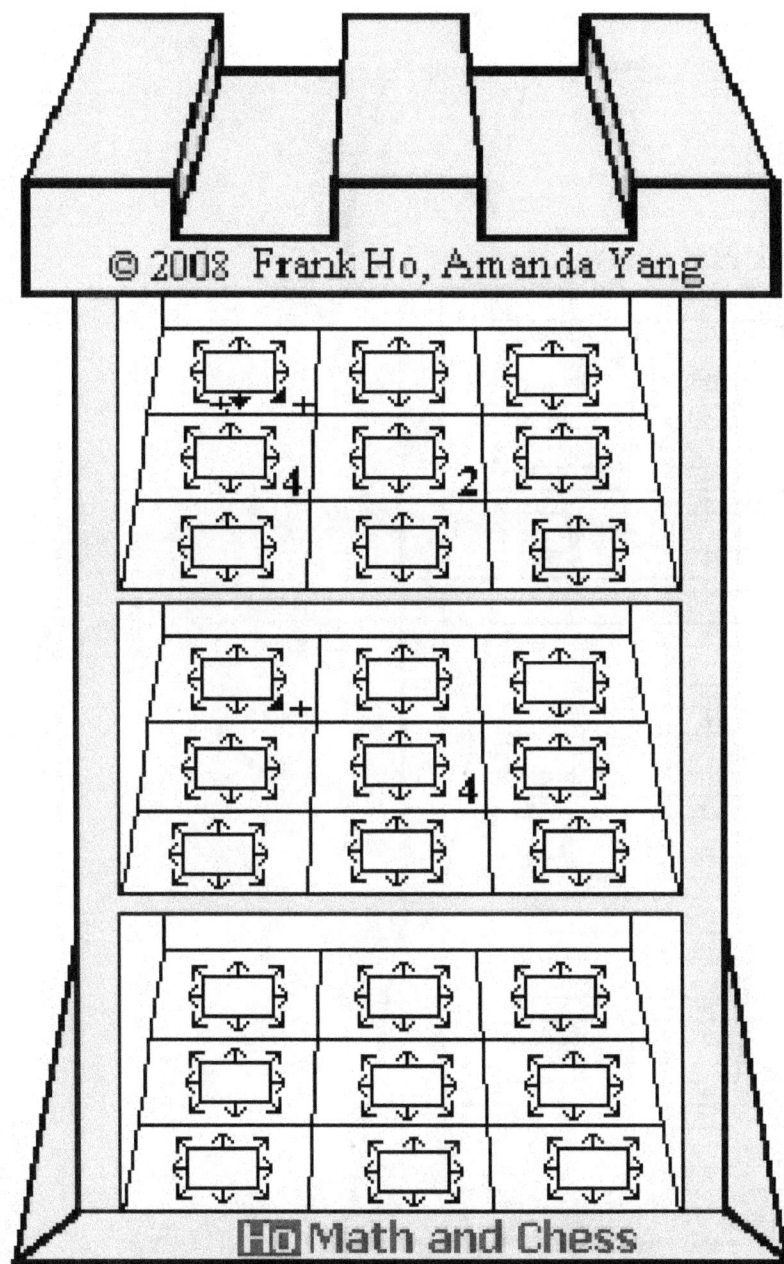

© 2008　Frank Ho, Amanda Yang

Ho Math and Chess

3 Dimensional Frankho ChessDoku™ # 3

Rule: All the digits 1 to 3 must appear in every row and column but cannot repeat on the same row of the same column of each layer. The number appears in the bottom right-hand corner is the end result calculated according to operator(s) and chess move(s).

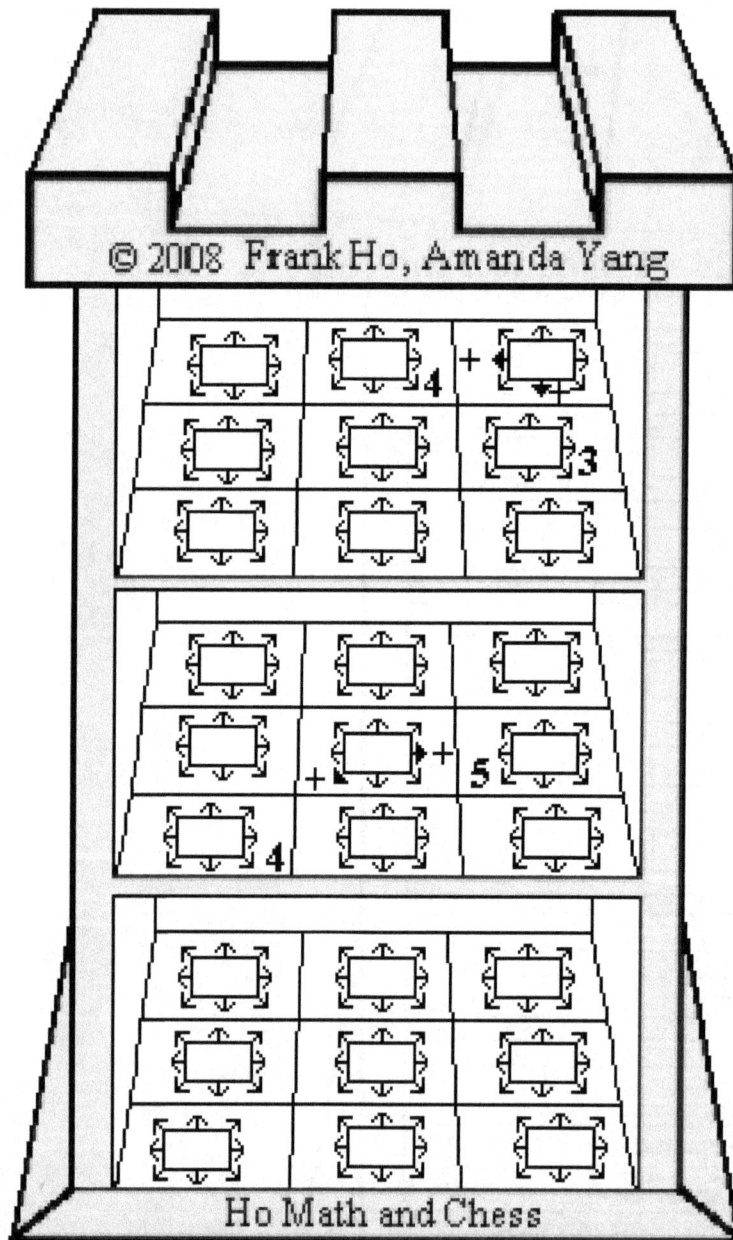

© 2008 Frank Ho, Amanda Yang

Ho Math and Chess

3 Dimensional Frankho ChessDoku™ # 4

Rule: All the digits 1 to 3 must appear in every row and column but cannot repeat on the same row of the same column of each layer. The number appears in the bottom right-hand corner is the end result calculated according to operator(s) and chess move(s).

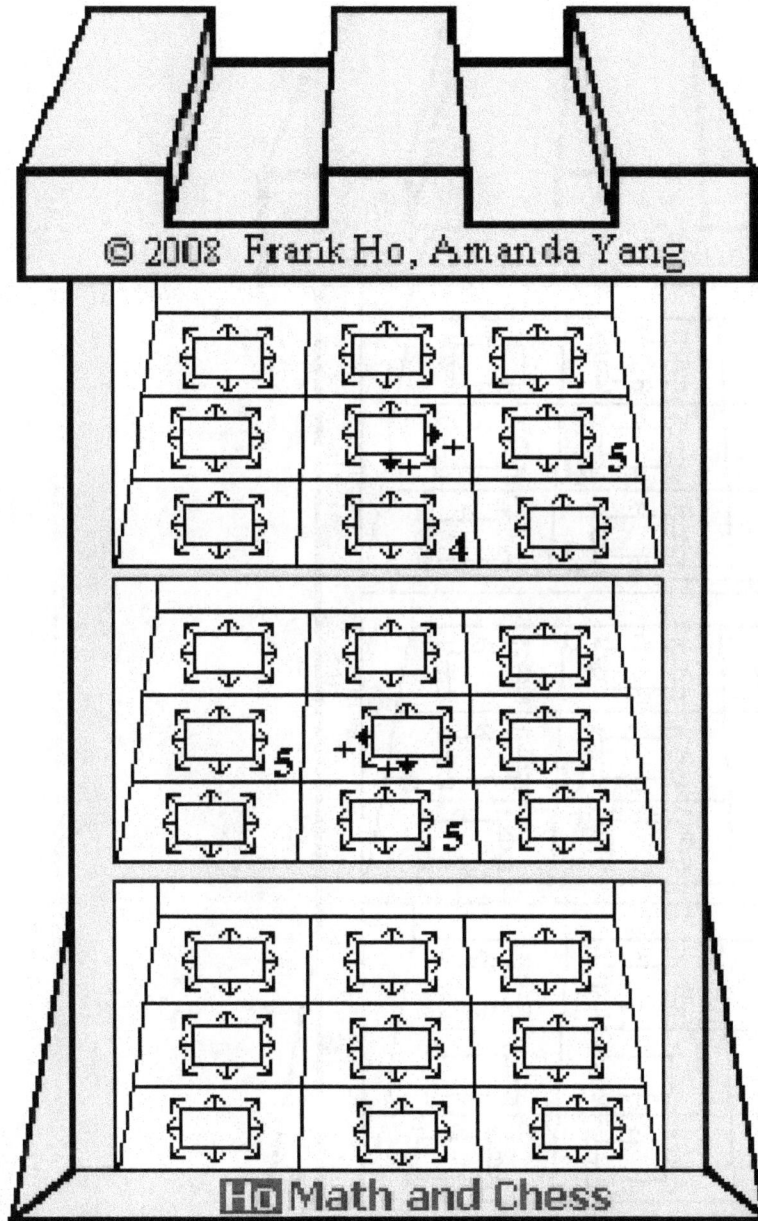

© 2008 Frank Ho, Amanda Yang

Ho Math and Chess

3 Dimensional Frankho ChessDoku™ # 5

Rule: All the digits 1 to 3 must appear in every row and column but cannot repeat on the same row of the same column of each layer. The number appears in the bottom right-hand corner is the end result calculated according to operator(s) and chess move(s).

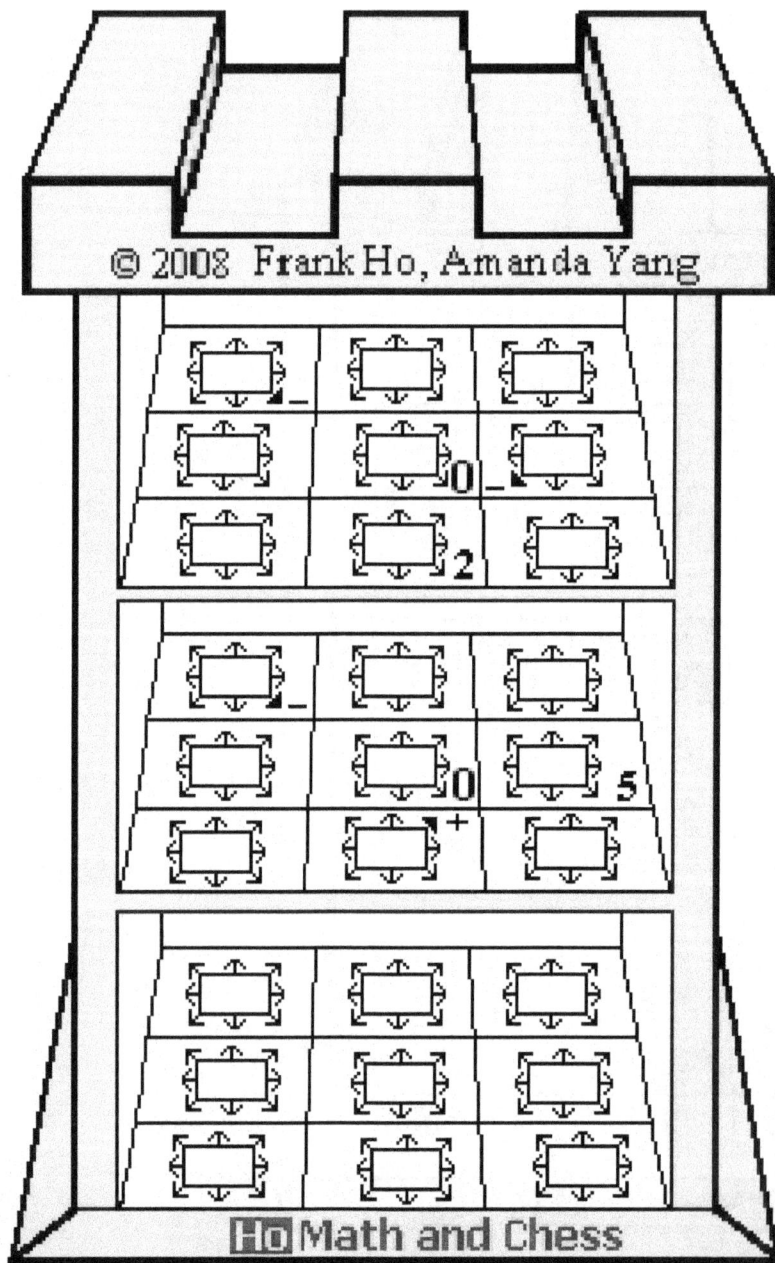

© 2008 Frank Ho, Amanda Yang

3 Dimensional Frankho ChessDoku™ # 6

Rule: All the digits 1 to 3 must appear in every row and column but cannot repeat on the same row of the same column of each layer. The number appears in the bottom right-hand corner is the end result calculated according to operator(s) and chess move(s).

© 2008 Frank Ho, Amanda Yang

Ho Math and Chess

Student's name _____ Date _____

3 Dimensional Frankho ChessDoku™ # 7

Rule: All the digits 1 to 3 must appear in every row and column but cannot repeat on the same row of the same column of each layer. The number appears in the bottom right-hand corner is the end result calculated according to operator(s) and chess move(s).

© 2008 Frank Ho, Amanda Yang

Student's name _____ Date _____

3 Dimensional Frankho ChessDoku™ # 8

Rule: All the digits 1 to 3 must appear in every row and column but cannot repeat on the same row of the same column of each layer. The number appears in the bottom right-hand corner is the end result calculated according to operator(s) and chess move(s).

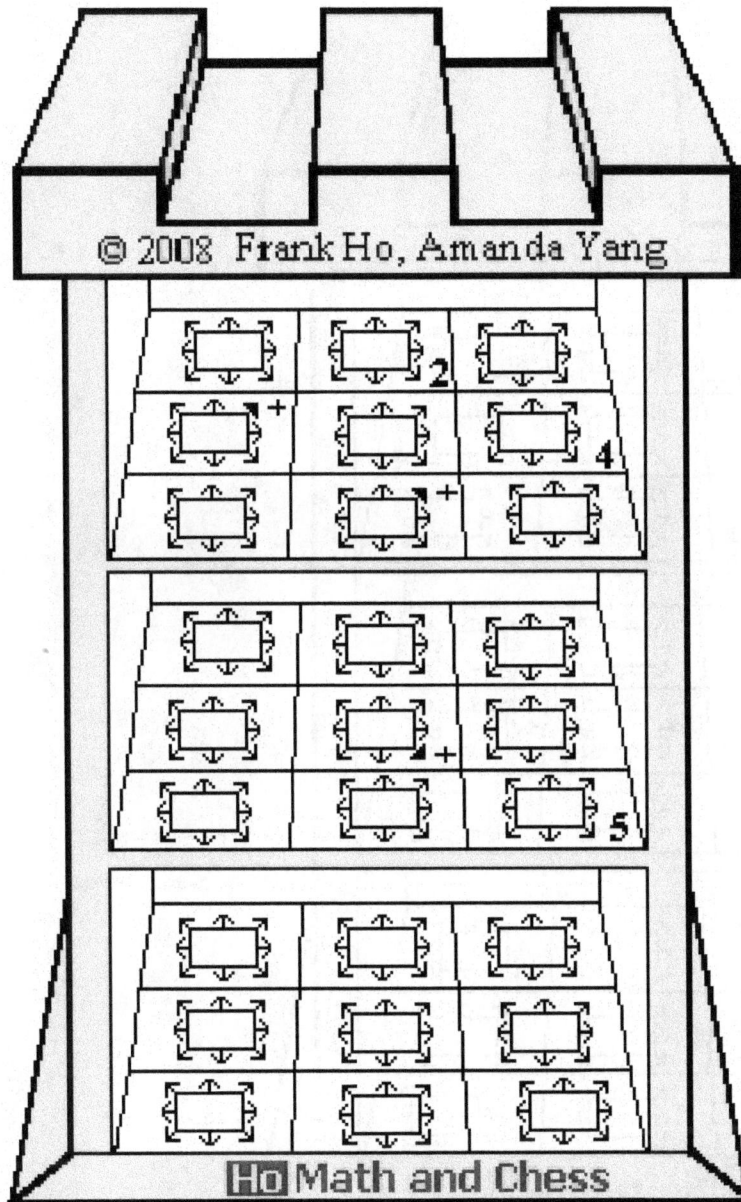

© 2008 Frank Ho, Amanda Yang

Ho Math and Chess

3 Dimensional Frankho ChessDoku™ # 9

Rule: All the digits 1 to 3 must appear in every row and column but cannot repeat on the same row
of the same column of each layer. The number appears in the bottom right-hand corner is
the end result calculated according to operator(s) and chess move(s).

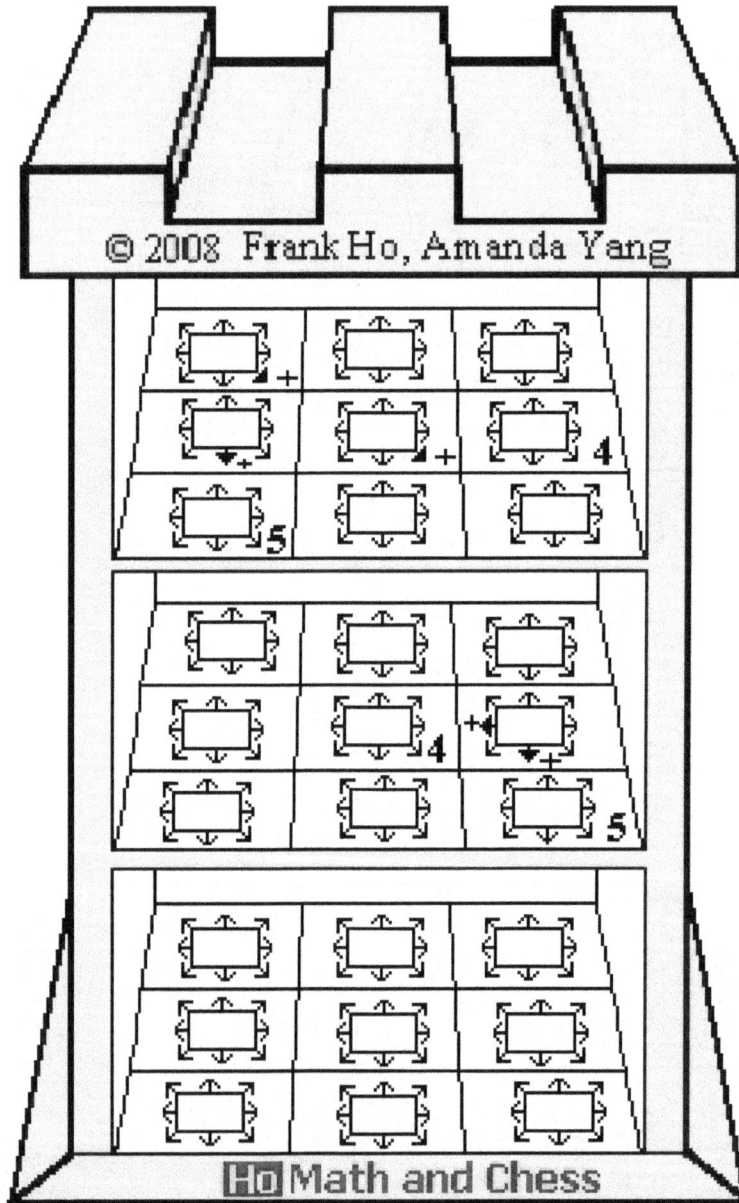

3 Dimensional Frankho ChessDoku™ # 10

Rule: All the digits 1 to 3 must appear in every row and column but cannot repeat on the same row of the same column of each layer. The number appears in the bottom right-hand corner is the end result calculated according to operator(s) and chess move(s).

© 2008 Frank Ho, Amanda Yang

Ho Math and Chess

167

3 Dimensional Frankho ChessDoku™ # 11

Rule: All the digits 1 to 3 must appear in every row and column but cannot repeat on the same row of the same column of each layer. The number appears in the bottom right-hand corner is the end result calculated according to operator(s) and chess move(s).

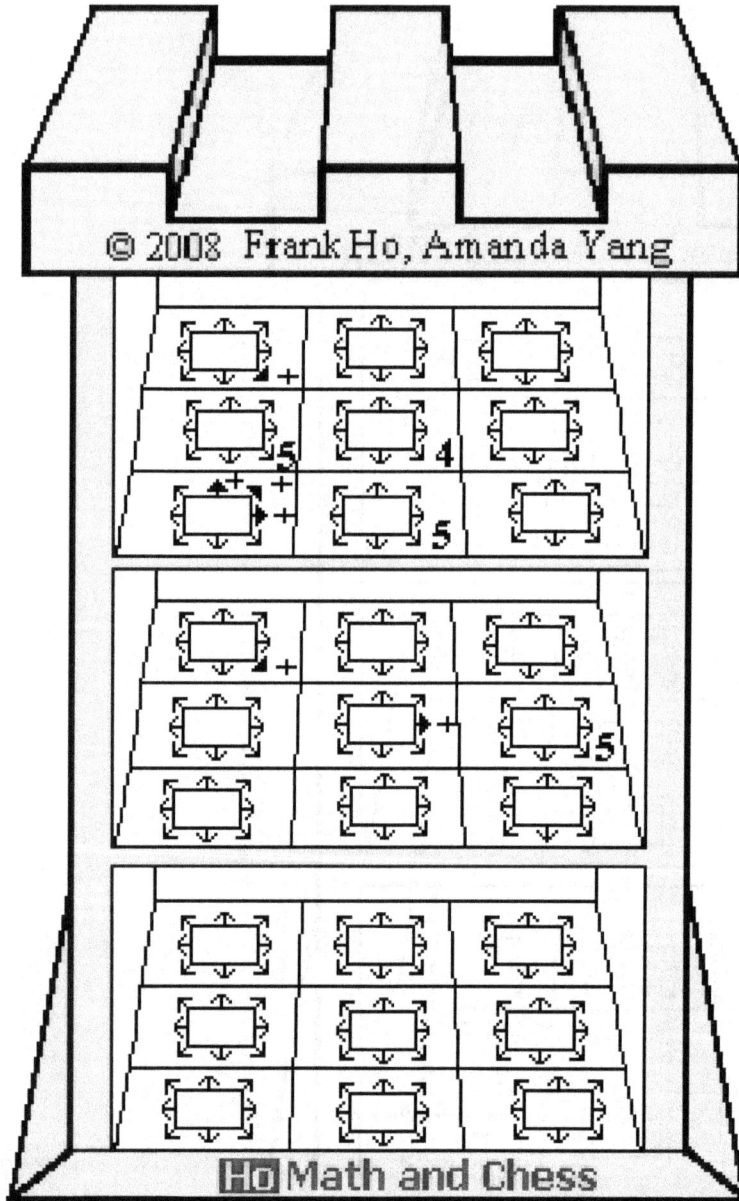

© 2008 Frank Ho, Amanda Yang

Ho Math and Chess

Student's name _____ Date _____

3 Dimensional Frankho ChessDoku™ # 12

Rule: All the digits 1 to 3 must appear in every row and column but cannot repeat on the same row of the same column of each layer. The number appears in the bottom right-hand corner is the end result calculated according to operator(s) and chess move(s).

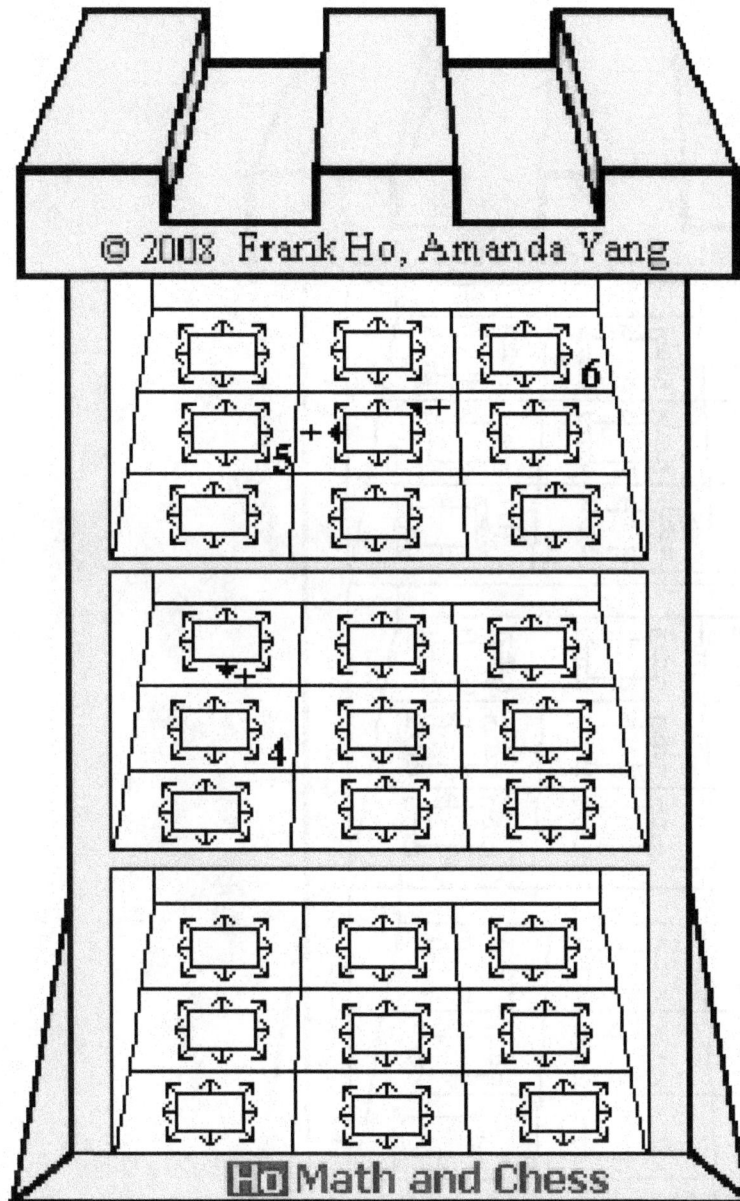

© 2008 Frank Ho, Amanda Yang

Ho Math and Chess

3 Dimensional Frankho ChessDoku™ # 13

Rule: All the digits 1 to 3 must appear in every row and column but cannot repeat on the same row of the same column of each layer. The number appears in the bottom right-hand corner is the end result calculated according to operator(s) and chess move(s).

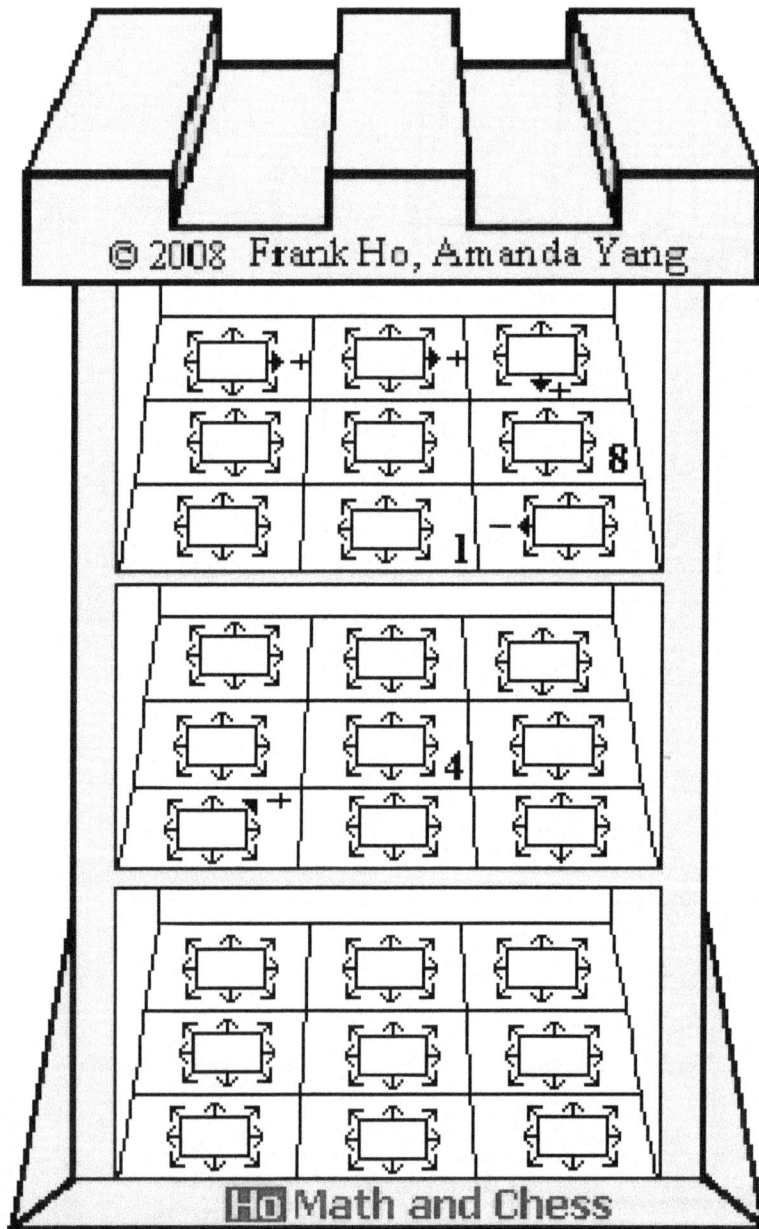

© 2008 Frank Ho, Amanda Yang

Ho Math and Chess

3 Dimensional Frankho ChessDoku™ # 14

Rule: All the digits 1 to 3 must appear in every row and column but cannot repeat on the same row of the same column of each layer. The number appears in the bottom right-hand corner is the end result calculated according to operator(s) and chess move(s).

© 2008 Frank Ho, Amanda Yang

Ho Math and Chess

3 Dimensional Frankho ChessDoku™ # 15

Rule: All the digits 1 to 3 must appear in every row and column but cannot repeat on the same row of the same column of each layer. The number appears in the bottom right-hand corner is the end result calculated according to operator(s) and chess move(s).

© 2008 Frank Ho, Amanda Yang

Ho Math and Chess

3 Dimensional Frankho ChessDoku™ # 16

Rule: All the digits 1 to 3 must appear in every row and column but cannot repeat on the same row
   of the same column of each layer. The number appears in the bottom right-hand corner is
   the end result calculated according to operator(s) and chess move(s).

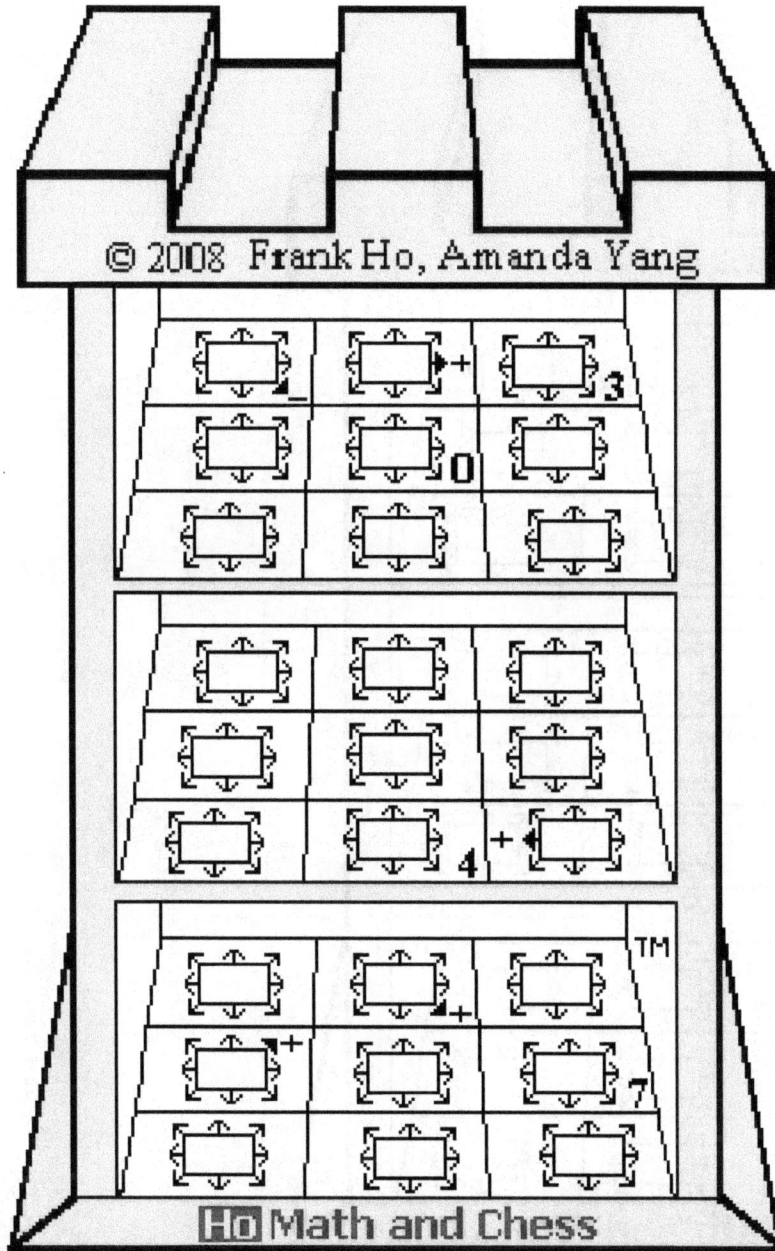

3 Dimensional Frankho ChessDoku™ # 17

Rule: All the digits 1 to 3 must appear in every row and column but cannot repeat on the same row of the same column of each layer. The number appears in the bottom right-hand corner is the end result calculated according to operator(s) and chess move(s).

© 2008 Frank Ho, Amanda Yang

Ho Math and Chess

Student's name _____ Date _____

3 Dimensional Frankho ChessDoku™ # 18

Rule: All the digits 1 to 3 must appear in every row and column but cannot repeat on the same row of the same column of each layer. The number appears in the bottom right-hand corner is the end result calculated according to operator(s) and chess move(s).

© 2008 Frank Ho, Amanda Yang

Ho Math and Chess

3 Dimensional Frankho ChessDoku™ # 19

Rule: All the digits 1 to 3 must appear in every row and column but cannot repeat on the same row of the same column of each layer. The number appears in the bottom right-hand corner is the end result calculated according to operator(s) and chess move(s).

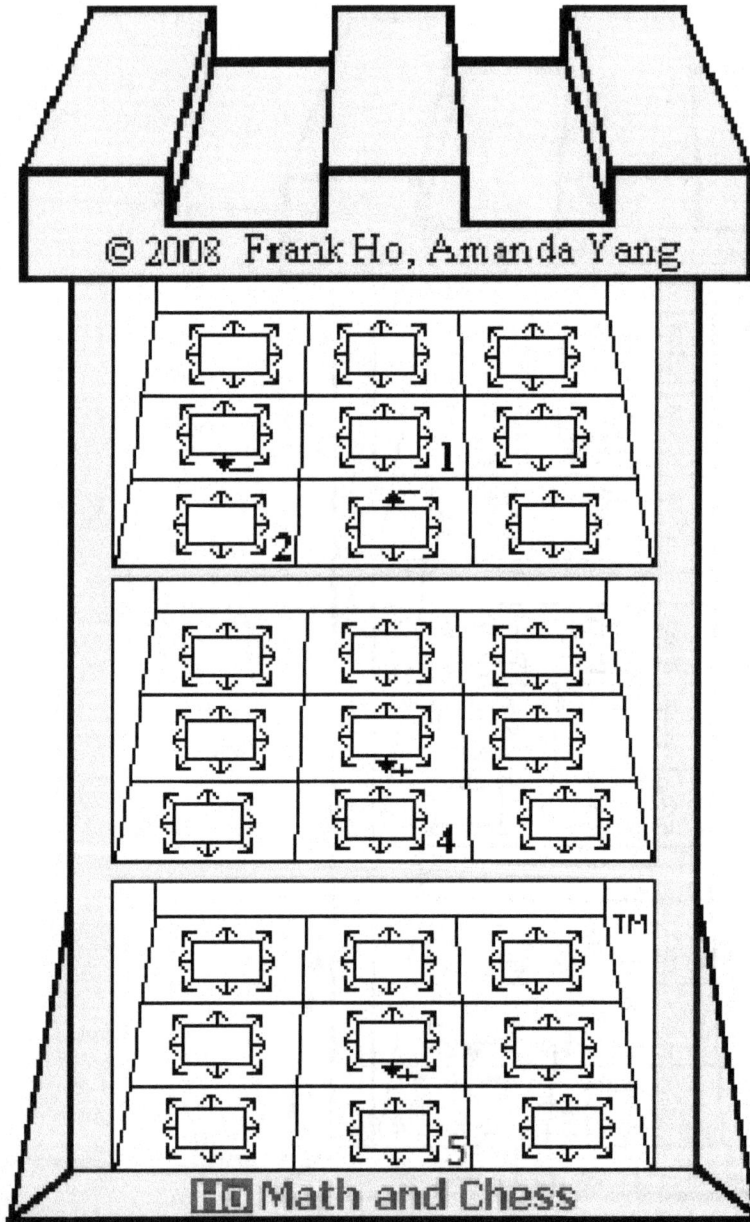

© 2008 Frank Ho, Amanda Yang

Ho Math and Chess

Student's name _____ Date _____

3 Dimensional Frankho ChessDoku™ # 20

Rule: All the digits 1 to 3 must appear in every row and column but cannot repeat on the same row of the same column of each layer. The number appears in the bottom right-hand corner is the end result calculated according to operator(s) and chess move(s).

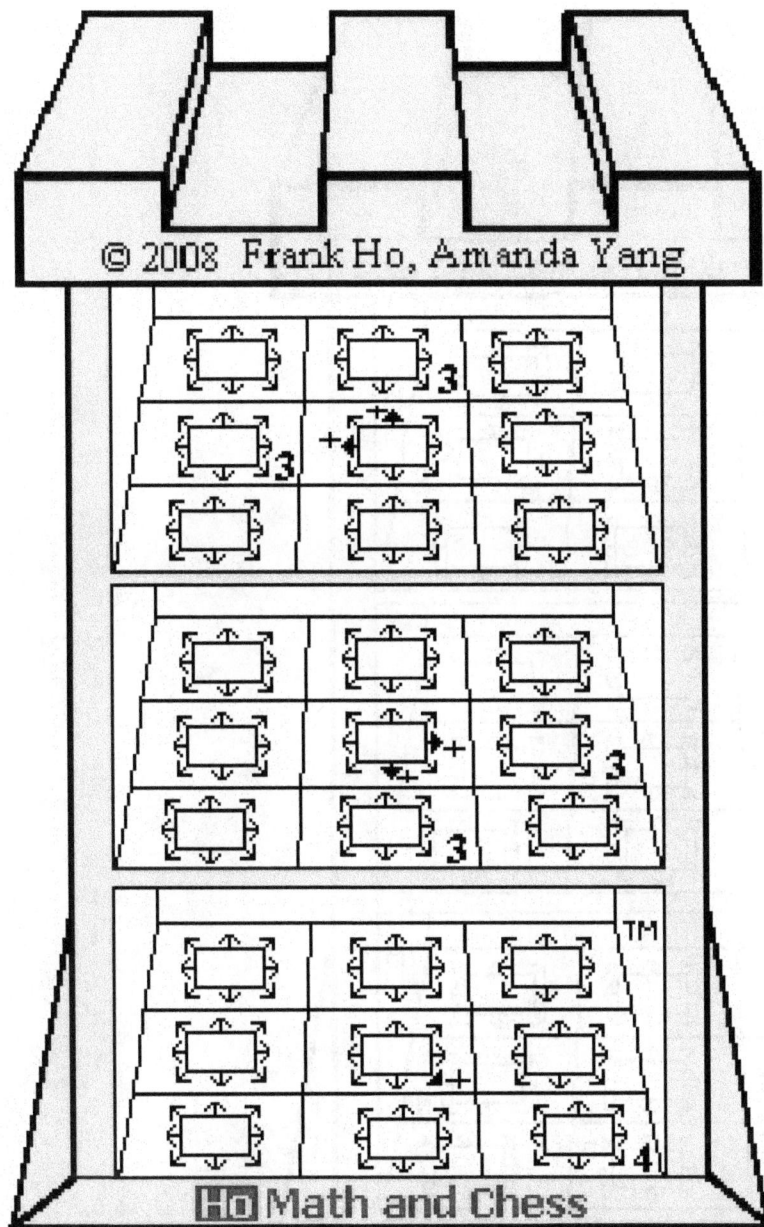

© 2008 Frank Ho, Amanda Yang

3 Dimensional Frankho ChessDoku™ # 21

Rule: All the digits 1 to 3 must appear in every row and column but cannot repeat on the same row of the same column of each layer. The number appears in the bottom right-hand corner is the end result calculated according to operator(s) and chess move(s).

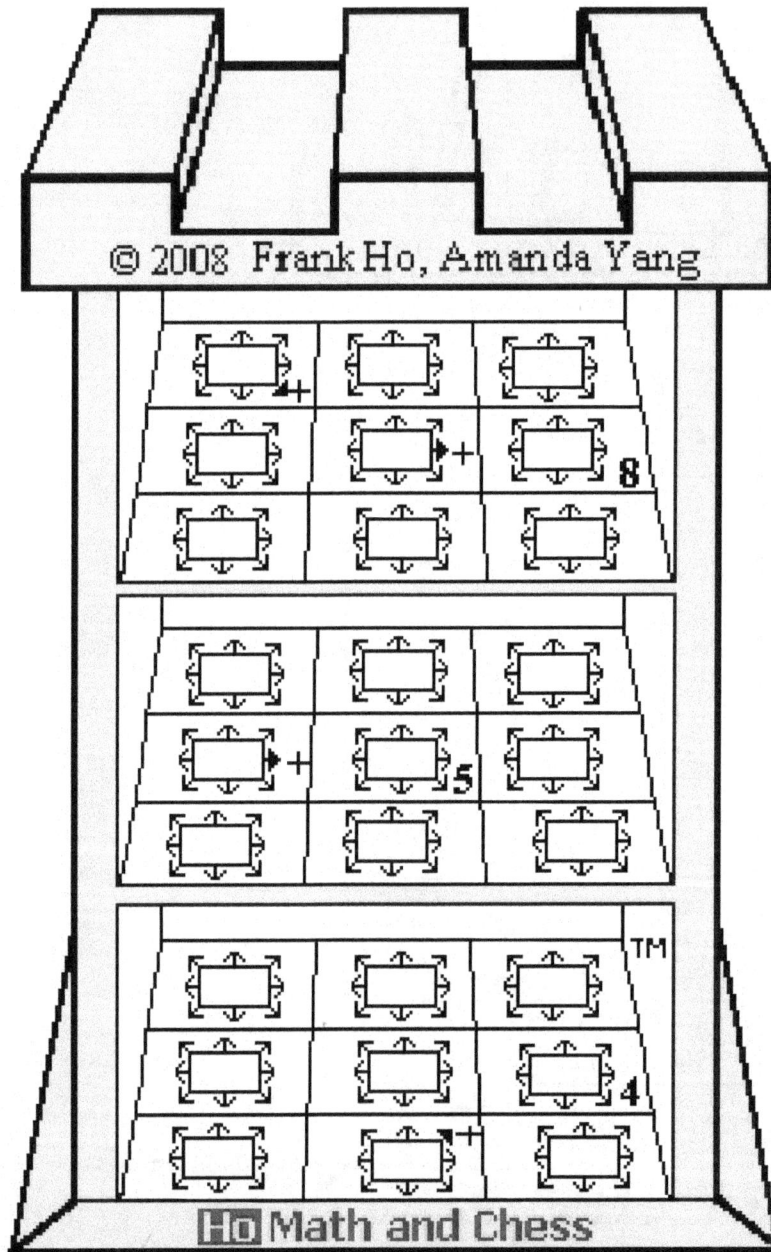

© 2008 Frank Ho, Amanda Yang

Ho Math and Chess

3 Dimensional Frankho ChessDoku™ # 22

Rule: All the digits 1 to 3 must appear in every row and column but cannot repeat on the same row of the same column of each layer. The number appears in the bottom right-hand corner is the end result calculated according to operator(s) and chess move(s).

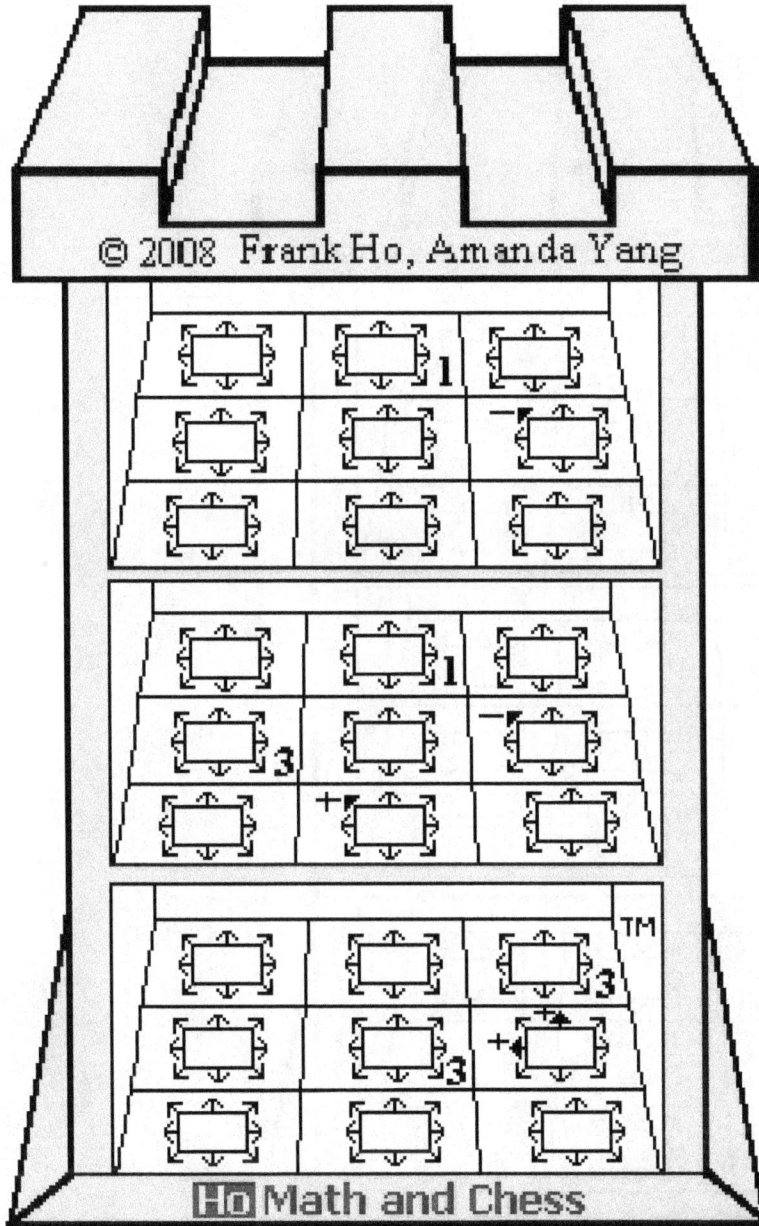

© 2008 Frank Ho, Amanda Yang

Ho Math and Chess

179

3 Dimensional Frankho ChessDoku™ # 23

Rule: All the digits 1 to 3 must appear in every row and column but cannot repeat on the same row of the same column of each layer. The number appears in the bottom right-hand corner is the end result calculated according to operator(s) and chess move(s).

© 2008　Frank Ho, Amanda Yang

Ho Math and Chess

Student's name _____ Date _____

3 Dimensional Frankho ChessDoku™ # 24

Rule: All the digits 1 to 3 must appear in every row and column but cannot repeat on the same row of the same column of each layer. The number appears in the bottom right-hand corner is the end result calculated according to operator(s) and chess move(s).

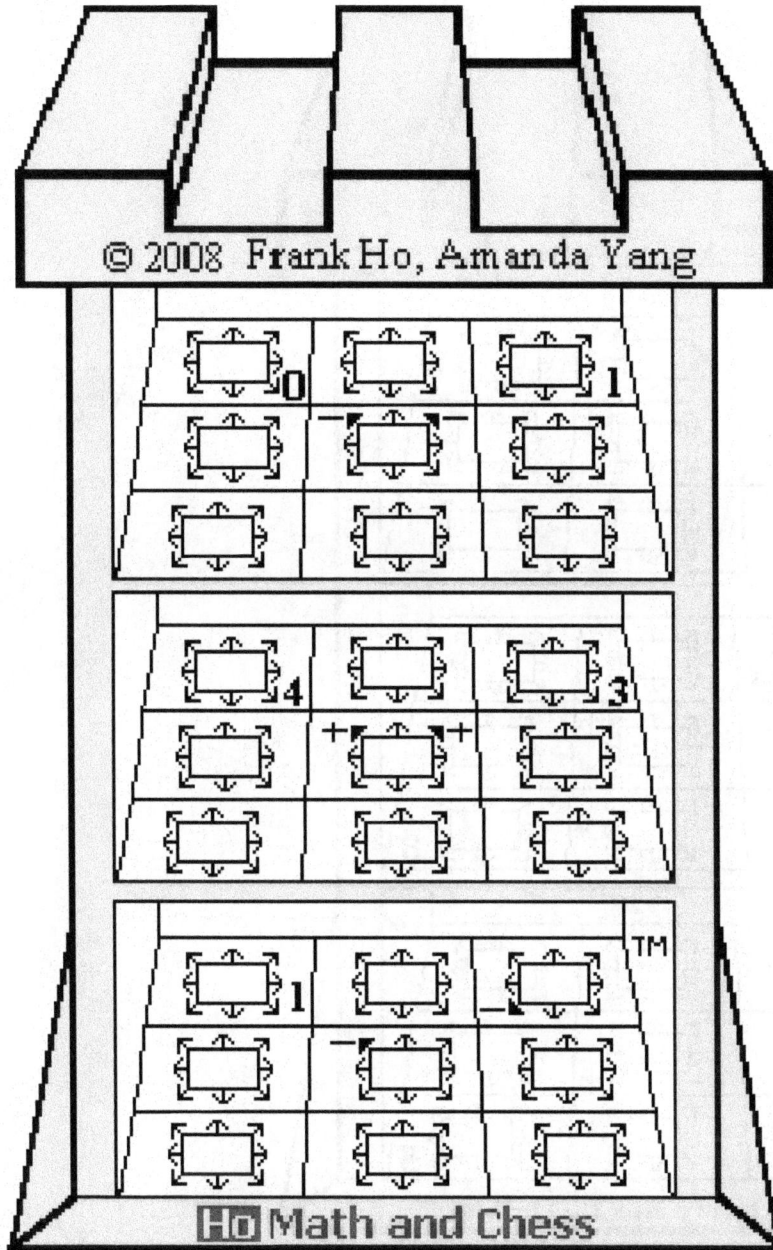

3 Dimensional Frankho ChessDoku™ # 25

Rule: All the digits 1 to 3 must appear in every row and column but cannot repeat on the same row of the same column of each layer. The number appears in the bottom right-hand corner is the end result calculated according to operator(s) and chess move(s).

© 2008 Frank Ho, Amanda Yang

Ho Math and Chess

3 Dimensional Frankho ChessDoku™ # 26

Rule: All the digits 1 to 3 must appear in every row and column but cannot repeat on the same row of the same column of each layer. The number appears in the bottom right-hand corner is the end result calculated according to operator(s) and chess move(s).

Student's name _____ Date _____

3 Dimensional Frankho ChessDoku™ # 27

Rule: All the digits 1 to 3 must appear in every row and column but cannot repeat on the same row of the same column of each layer. The number appears in the bottom right-hand corner is the end result calculated according to operator(s) and chess move(s).

© 2009 Frank Ho, Amanda Yang

Ho Math and Chess

3 Dimensional Frankho ChessDoku™ # 28

Rule: All the digits 1 to 3 must appear in every row and column but cannot repeat on the same row
of the same column of each layer. The number appears in the bottom right-hand corner is
the end result calculated according to operator(s) and chess move(s).

© 2009 Frank Ho, Amanda Yang

Ho Math and Chess

3 Dimensional Frankho ChessDoku™ # 29

Rule: All the digits 1 to 3 must appear in every row and column but cannot repeat on the same row of the same column of each layer. The number appears in the bottom right-hand corner is the end result calculated according to operator(s) and chess move(s).

3 Dimensional Frankho ChessDoku™ # 30

Rule: All the digits 1 to 3 must appear in every row and column but cannot repeat on the same row of the same column of each layer. The number appears in the bottom right-hand corner is the end result calculated according to operator(s) and chess move(s).

© 2009 Frank Ho, Amanda Yang

Ho Math and Chess

3 Dimensional Frankho ChessDoku™ # 31

Rule: All the digits 1 to 3 must appear in every row and column but cannot repeat on the same row of the same column of each layer. The number appears in the bottom right-hand corner is the end result calculated according to operator(s) and chess move(s).

© 2009 Frank Ho, Amanda Yang

Ho Math and Chess

Student's name _____ Date _____

3 Dimensional Frankho ChessDoku™ # 32

Rule: All the digits 1 to 3 must appear in every row and column but cannot repeat on the same row of the same column of each layer. The number appears in the bottom right-hand corner is the end result calculated according to operator(s) and chess move(s).

© 2009 Frank Ho, Amanda Yang

Ho Math and Chess

3 Dimensional Frankho ChessDoku™ # 33

Rule: All the digits 1 to 3 must appear in every row and column but cannot repeat on the same row of the same column of each layer. The number appears in the bottom right-hand corner is the end result calculated according to operator(s) and chess move(s).

© 2009 Frank Ho, Amanda Yang

Ho Math and Chess

3 Dimensional Frankho ChessDoku™ # 34

Rule: All the digits 1 to 3 must appear in every row and column but cannot repeat on the same row of the same column of each layer. The number appears in the bottom right-hand corner is the end result calculated according to operator(s) and chess move(s).

1

3 Dimensional Frankho ChessDoku™ # 35

Rule: All the digits 1 to 3 must appear in every row and column but cannot repeat on the same row of the same column of each layer. The number appears in the bottom right-hand corner is the end result calculated according to operator(s) and chess move(s).

© 2009   Frank Ho, Amanda Yang

Ho Math and Chess

3

3 Dimensional Frankho ChessDoku™ # 36

Rule: All the digits 1 to 3 must appear in every row and column but cannot repeat on the same row of the same column of each layer. The number appears in the bottom right-hand corner is the end result calculated according to operator(s) and chess move(s).

© 2009 Frank Ho, Amanda Yang

Ho Math and Chess

3

3 Dimensional Frankho ChessDoku™ # 37

Rule: All the digits 1 to 3 must appear in every row and column but cannot repeat on the same row of the same column of each layer. The number appears in the bottom right-hand corner is the end result calculated according to operator(s) and chess move(s).

© 2009 Frank Ho, Amanda Yang

Ho Math and Chess

Student's name _____ Date _____

3 Dimensional Frankho ChessDoku™ # 38

Rule: All the digits 1 to 3 must appear in every row and column but cannot repeat on the same row of the same column of each layer. The number appears in the bottom right-hand corner is the end result calculated according to operator(s) and chess move(s).

© 2009 Frank Ho, Amanda Yang

Ho Math and Chess

3 Dimensional Frankho ChessDoku™ # 39

Rule: All the digits 1 to 3 must appear in every row and column but cannot repeat on the same row of the same column of each layer. The number appears in the bottom right-hand corner is the end result calculated according to operator(s) and chess move(s).

© 2009 Frank Ho, Amanda Yang

Ho Math and Chess

3 Dimensional Frankho ChessDoku™ # 40

Rule: All the digits 1 to 3 must appear in every row and column but cannot repeat on the same row of the same column of each layer. The number appears in the bottom right-hand corner is the end result calculated according to operator(s) and chess move(s).

3 Dimensional Frankho ChessDoku™ # 41

Rule: All the digits 1 to 3 must appear in every row and column but cannot repeat on the same row
   of the same column of each layer. The number appears in the bottom right-hand corner is
   the end result calculated according to operator(s) and chess move(s).

© 2009 Frank Ho, Amanda Yang

Ho Math and Chess

3 Dimensional Frankho ChessDoku™ # 42

Rule: All the digits 1 to 3 must appear in every row and column but cannot repeat on the same row of the same column of each layer. The number appears in the bottom right-hand corner is the end result calculated according to operator(s) and chess move(s).

3 Dimensional Frankho ChessDoku™ # 43

Rule: All the digits 1 to 3 must appear in every row and column but cannot repeat on the same row of the same column of each layer. The number appears in the bottom right-hand corner is the end result calculated according to operator(s) and chess move(s).

© 2009   Frank Ho,   Amanda Yang

Ho Math and Chess

3 Dimensional Frankho ChessDoku™ # 44

Rule: All the digits 1 to 3 must appear in every row and column but cannot repeat on the same row
  of the same column of each layer. The number appears in the bottom right-hand corner is
  the end result calculated according to operator(s) and chess move(s).

© 2009 Frank Ho, Amanda Yang

Ho Math and Chess

3 Dimensional Frankho ChessDoku™ # 45

Rule: All the digits 1 to 3 must appear in every row and column but cannot repeat on the same row of the same column of each layer. The number appears in the bottom right-hand corner is the end result calculated according to operator(s) and chess move(s).

© 2009 Frank Ho, Amanda Yang

Ho Math and Chess

3 Dimensional Frankho ChessDoku™ # 46

Rule: All the digits 1 to 3 must appear in every row and column but cannot repeat on the same row of the same column of each layer. The number appears in the bottom right-hand corner is the end result calculated according to operator(s) and chess move(s).

3 Dimensional Frankho ChessDoku™ # 47

Rule: All the digits 1 to 3 must appear in every row and column but cannot repeat on the same row of the same column of each layer. The number appears in the bottom right-hand corner is the end result calculated according to operator(s) and chess move(s).

3 Dimensional Frankho ChessDoku™ # 48

Rule: All the digits 1 to 3 must appear in every row and column but cannot repeat on the same row
of the same column of each layer. The number appears in the bottom right-hand corner is
the end result calculated according to operator(s) and chess move(s).

3 Dimensional Frankho ChessDoku™ # 49

Rule: All the digits 1 to 3 must appear in every row and column but cannot repeat on the same row
   of the same column of each layer. The number appears in the bottom right-hand corner is
   the end result calculated according to operator(s) and chess move(s).

© 2009 Frank Ho, Amanda Yang

Ho Math and Chess

3 Dimensional Frankho ChessDoku™ # 50

Rule: All the digits 1 to 3 must appear exactly once in every row and column but cannot repeat on the same row or the same column of each layer. The number appears in the bottom right-hand corner is the end result calculated according to operator(s) and chess move(s) as indicated by darker arrow(s).

3 Dimensional Frankho ChessDoku™ # 51

Rule: All the digits 4 to 6 must appear exactly once in every row and column but cannot repeat on the same row or the same column of each layer. The number appears in the bottom right-hand corner is the end result calculated according to operator(s) and chess move(s) as indicated by darker arrow(s).

© 2009 Frank Ho, Amanda Yang

Ho Math and Chess

Comparing Frankho ChessDoku and CalcuDoku

何数棋算独与普通算独的比较
何数棋算独是由加拿大何老师发明以数独游戏与数学运算及国际象棋的巧妙结合
Frank Ho, Amanda Ho

April, 2013

Introduction of Frankho ChessDoku ©

Frankho ChessDoku was invented by a Canadian math teacher Frank Ho (1, 2). Seeing the popularity of Sudoku but with no computation capability, Frank decided to do something about it so Frank used his invented Geometry Chess Symbols (Canada Trademark TMA771400, copyright 1069744) along with Sudoku created the *Frankho ChessDoku* in 2008. *Frankho ChessDoku* is a unique puzzle which combines arithmetic, chess, and Sudoku all in one puzzle and is specially designed for children to solve arithmetic using backwards strategy by following chess moves and logic. In 2009 Frank Ho and his wife Amanda Ho jointly published a workbook.

Frank always has an idea about teaching math that is students should always be encouraged and more importantly be given a chance to THINK, and it means even when they are doing pure computation problems. This is the reason he has created many basic number facts computation workbooks using the idea of integrating math, chess, and puzzles. Math + *Chess* + *Sudoku* = *Fun Frankho ChessDoku* ©

The pleasure of working on Ho Math Chess workbooks could be very well described by a famous classical Chinese poem 山重水复疑无路,柳暗花明又一村 (Equivalent English phrase is *seeing light at the end of tunnel.*)

Frank has described the characteristics of Ho Math Chess worksheets in Chinese rhyming sentences (打油诗) as follows. Its meaning mainly describes the magic of Ho Math Chess puzzles.

只见棋谜不见题　　劝君迷路不哭涕
数学象棋加谜题　　健脑思维真神奇

Introduction of CalcuDoku

The original CalcuDoku was invented in 2004 by a Japanese teacher Tetsuya Miyamoto in Japan (3).

Comparisons

The key difference between *Frankho ChessDoku* and CalcuDoku is that *Frankho* ChessDoku uses Frank's invented Geometry Chess Symbols to guide children on the directions of arithmetic operations instead of using "boxes or "cages" as used in Miyamoto's puzzles.

Frankho ChessDoku does not just use chess pieces to replace numbers in Sudoku as seen in some ChessDoku puzzles. *Frankho ChessDoku* invites children to trace chess moves to see the results just as if they were playing chess game by examining the intersections of chess moves and then use the logic of Sudoku to figure out the answers. Both strategies of playing chess game especially the intersections of chess moves and the arithmetic Sudoku logic need to be combined to solve *Frankho ChessDoku* puzzles.

Miyamoto runs a learning centre in Japan and teaches his puzzles to children. Frank and his wife also use their puzzles to teach children from age 4 and up in their Ho Math Chess learning centre in Vancouver, Canada. Both Frank and his wife teach children from kindergarten and up and both of them also teach math contest preparations.

From a student's learning math point of view, *Frankho ChessDoku* offers more powerful learning and mental training advantages over regular Sudoku and also other types of arithmetic Sudoku. The following table gives some comparisons. In addition to being a fun puzzle, *Frankho ChessDoku* is more suitable for students who like to improve their brainpower and also mental math ability.

	Frankho ChessDoku	Regular Sudoku	CalcuDoku
addition, subtraction, multiplication , and division	Can provide 4 mixed operations by following chess moves within one equation with no confusion.	No computations	Only independent and separate +, −, ×, ÷ operations can be provided. Mixed 4 basic operations could cause confusion for young children.
vertical, horizontal, and diagonal operations	The horizontal or vertical operations are provided. The diagonal operations can also be provided. The "jump" operation (knight move) can be provided.	No computations	Only horizontal or vertical operations are provided. No diagonal operations can be provided. No "jump" operation can be provided.
Framed or boxed operations	No framed operation is required since the operation direction is guided by chess moves. Children can always circle operation statements themselves. This flexibility allows intersecting "boxes" with no confusion.	No computations	Since operation is always "boxed" or "caged" with single operation, so no flexibility is allowed for mixed operations. Intersecting frames or boxes would cause confusion.

Student's name _____Date _____

Example 1

Rule

All the digits 1 to 3 must appear exactly once in every row and column. The number appears in the bottom right-hand corner is the end result calculated according to arithmetic operator(s) and chess move(s) as indicated by darker arrow(s).

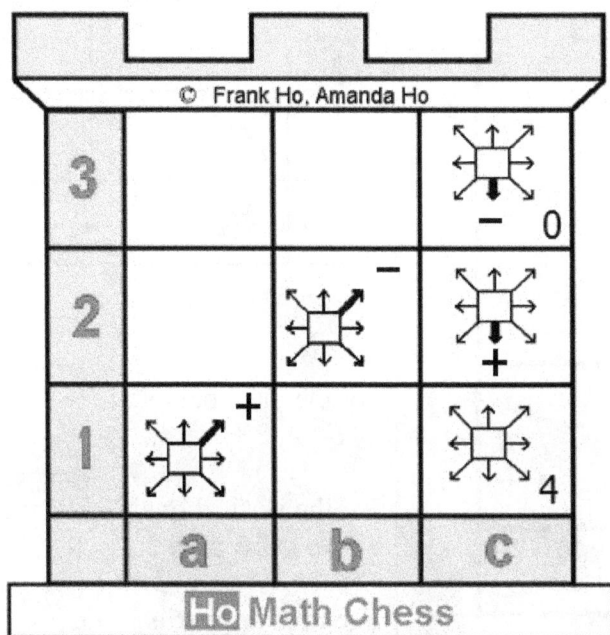

A CalcuDoku is not able to produce the diagonal operation of the above *Frankho ChessDoku*.

The mixed operation $3 + 2 - 1 = 4$ or $2 + 3 - 1 = 4$ is also difficult for children to work on if it happens in the CalcuDoku since it involves two operations at the same time but is very easy for *Frankho ChessDoku* to identify it with no confusion. This deficiency in CalcuDoku means children seem to always be stuck with only one operation at a time with very little chance to work on mixed operations. In contrast, Children working on *Frankho ChessDoku* will have plenty of opportunities to work on either single or mixed operations with no confusion simply by following chess moves.

Step 1:
Circle all operations by following chess moves.

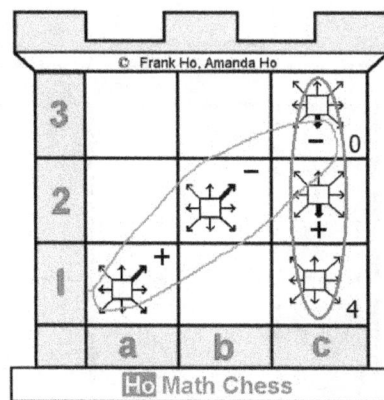

Step 2.
For the diagonal oval, $2 + 1 - 3 = 0$,
$1 + 2 - 3 = 0$.

For the vertical oval, $3 - 1 + 2 = 4$
So we know c3=3.

The final answer is as follows.

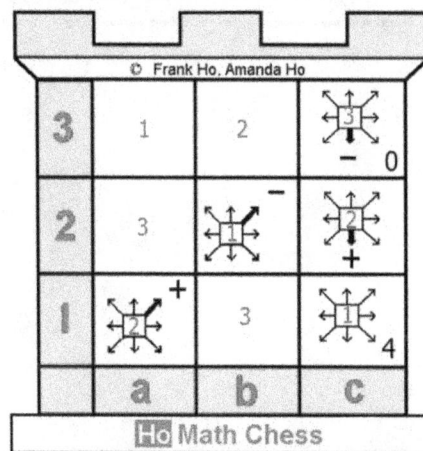

211

Example 2

A CalcuDoku cannot produce the following "Jump" movement as acting by the chess knight move at c3.

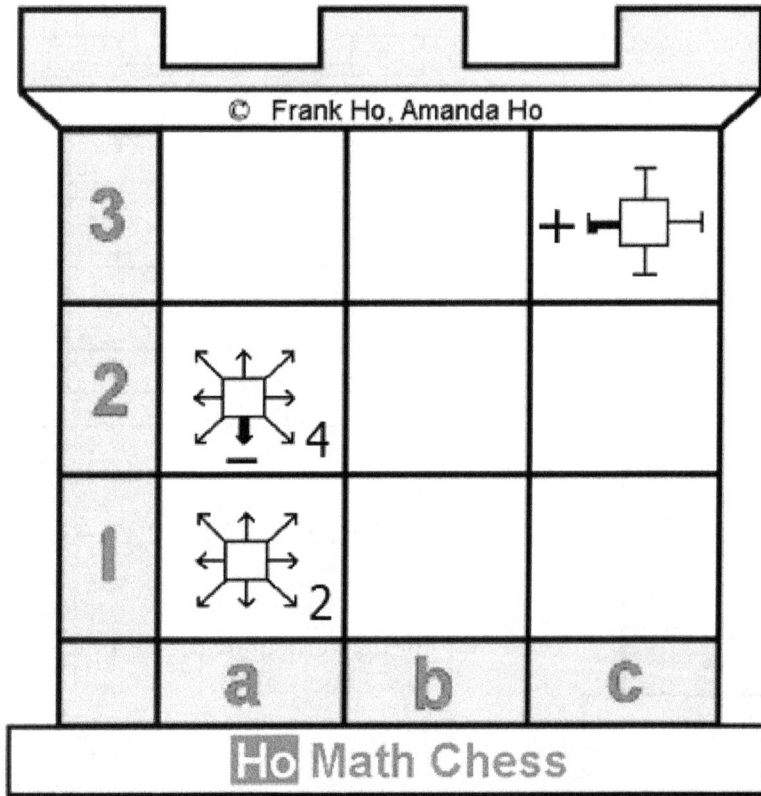

Step 1

Circle all operations.

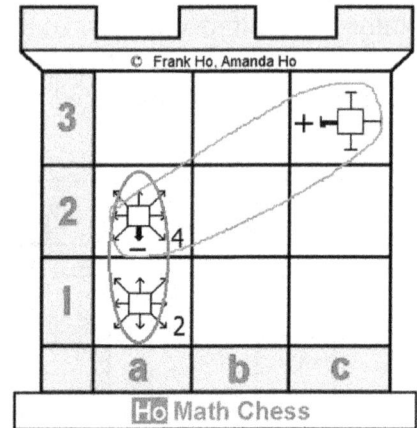

Step 2

For the diagonal oval, $1 + 3 = 4$, $3 + 1 = 4$, $2 + 2 = 4$.

For the vertical oval, $3 - 1 = 2$. So we know a2 = 3.

The final answer is

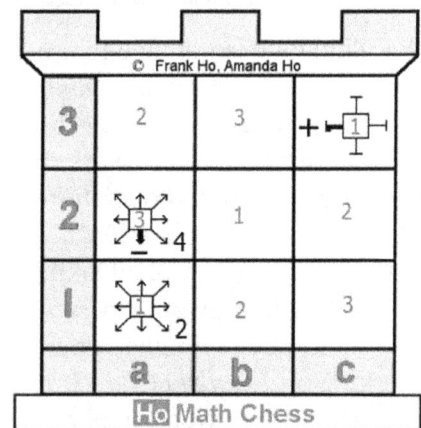

Example 3

The famous Sum and Difference problem can be easily illustrated by using the above *Frankho ChessDoku diagram* with intersection and also can be very easily solved, but trying to create it using the idea of CalcuDoku demonstrates confusion for children.

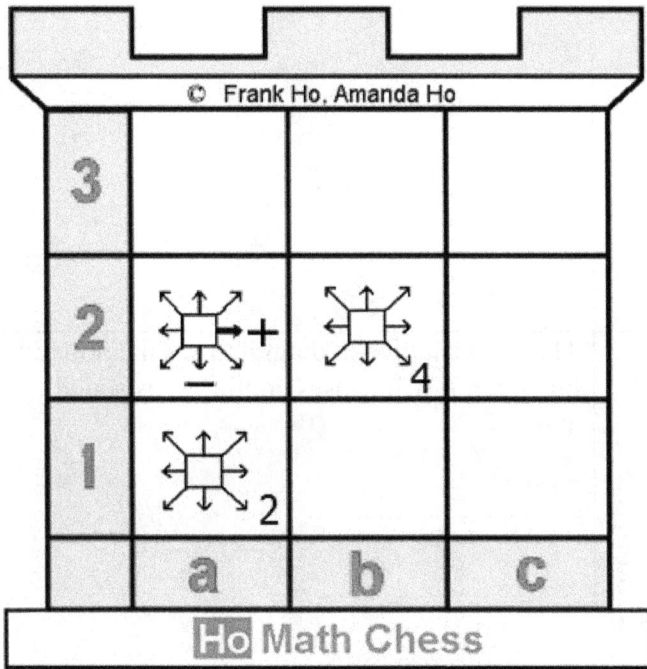

Step 1

Circle all operations.

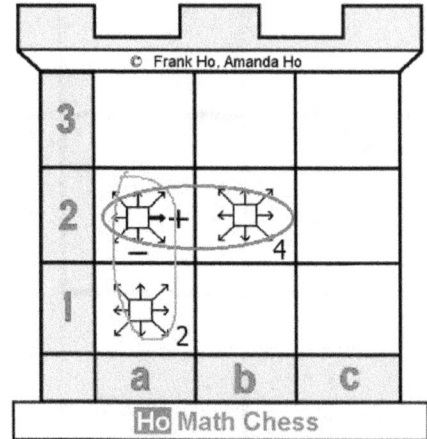

Step 2

Start at intersection a2.
For the horizontal oval, 3 + 1 = 4,
1+ 3 = 4, 2 + 2 = 4.
For the vertical oval, 3 − 1 = 2
So we know a2 = 3.

The final answer is as follows:

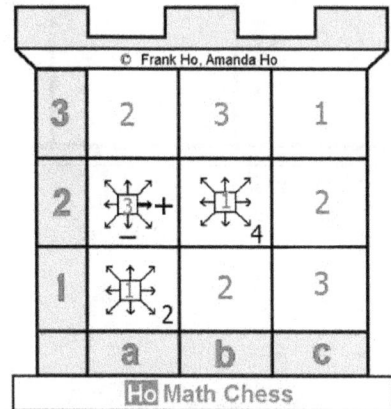

The following are the same problem (Sum and Difference) using the diagrams of CalcuDoku.

The left CalcuDoku diagram causes confusion because we do not know which box is for 4 + and which box is for 1−. The Venn diagram concept can be easily demonstrated in the *Frankho ChessDoku* but causes confusion in the CalcuDoku.

The left CalcuDoku diagram uses the dotted box but again it still causes confusion as stated above.

Commutative law

The convention way of calculating is in the direction of left to right or top to down, but this rule does not apply to CalcuDoku because as shown below" 2 – " can be expressed as 3, 1 or 1, 3 and it appears to students that the subtraction can be done by exchanging the two numbers and this is in violation of the commutative law. It would have no problem for *Frankho ChessDoku* to handle the subtraction and division operators because the calculation direction is clearly defined by using chess moves.

For subtraction operator, the answer could be operated from left to right but sometimes, it could be also from right to left.

The above CalcuDoku does require student to think about how 3 and 1 are to be arranged, so it could present extra challenging for students, but the confusion could also occur when the mixed operators (+, −, ×, ÷) are presented together with no operating directions are given.

The left *Frankho ChessDoku* presents no operation confusion and does not have the confusion of commutative law for children.

215

Chess strategy and *Frankho ChessDoku* strategy

Often a chess player would analyze the chess moves and see where each chess piece intersects each other, then decide to take the action of next move. This kind of thinking is also reflected in the strategy on how to solve *Frankho ChessDoku* and the following example demonstrated the transferred knowledge between chess and *Frankho ChessDoku*.

Find a Black move to fork.
Qe1 moves to f2 to fork black king and knight.

The above c1 intersects with b2 and c2 at the same time so in other words, c1 is a square where bishop and rook intersects in chess. This kind of thinking is no difference from the chess diagram on the left to consider at what square where the queen could move to such that the queen could fork Black king and knight at the same time.

Frankho ChessDoku trains children to watch the intersections of lines and this knowledge could be transferred to benefitting chess play.

216

Triangular solving strategy for 3 by 3 grid

The simplest 3 by 3 case of *Frankho ChessDoku* can be created by using only one number and one math operator. All other math operations are really redundant.

The triangular method can be used to decide the number at b2 which must be 3.

The above diagonal operation cannot be duplicated by CalcuDoku; instead a horizontal operation could be made. The operation directions in CalcuDoku can be replicated in Frankho ChessDoku but not always the other way around. Frankho ChessDoku is much more flexible in terms of mixed calculations and also trains more visualization.

Frankho ChessDoku

References

(1) http://susanpolgar.blogspot.ca/2008/10/ho-math-and-chess.html
(2) http://www.mathandchess.com/releases/release/1441781/17465.htm
(3) http://en.wikipedia.org/wiki/Tetsuya_Miyamoto

Ho Math Chess 何数棋谜 益智健脑非药物良方
Frankho ChessDoku – Brain Fitness Workbook 何数棋算独

Student's name ＿＿＿＿＿＿＿＿＿＿＿＿＿＿＿＿＿Date ＿＿＿＿＿＿＿＿＿

Ho Math Chess other publications

Ho Math Chess has published many other workbooks for students to use. They are unique especially the basic number facts learning series because they use a fun and effective math, chess, and puzzles integrated methodology. At present, all these workbooks are only available through worldwide Ho Math Chess learning centre franchisees.

Pre-K and kindergarten Math
Kindergarten Math
Math Entrance Test Preparation for Primary Students
Learning Calculation Without Counting Fingers
Problems Solving and Math IQ Puzzles for Primary Grades
Frankho ChessDoku 3 by 3
Frankho ChessDoku 4 by 4 Volume 1
Frankho ChessDoku 4 by 4 Volume 2
Learning Chess to Improve Math
Mom! I Learn Addition Using Math-Chess-Puzzles connection
Mom! I Learn Subtraction Using Math-Chess-Puzzles connection
Mom! I Learn Addition and Subtraction Using Math-Chess-Puzzles connection
Mom! I Learn Division Using Math-Chess-Puzzles connection
Mom! I Learn Multiplication Using Math-Chess-Puzzles connection
Whole Number Operations
Fundamental Math
Math 8
Problem Solving and Math IQ Puzzles Volume 1
Problem Solving and Math IQ Puzzles Volume 2
Problem Solving and Math IQ Puzzles Volume 3
Problem Solving and Math IQ Puzzles Volume 4
Problem Solving and Math IQ Puzzles Volume 5
Ultimate Math Contests Preparation for Junior High School
Ultimate Math Contests Preparation for Beginners
Ultimate Math Contests Preparation for Intermediate Students
Ultimate Math Contests Preparation for Advanced Students Volume 1
Ultimate Math Contests Preparation for Advanced Students Volume 2
Why Buy a Ho Math Chess Learning Centre Franchise
Ho Math Chess Sudoku Puzzles Sample Worksheets
Introduction to Ho Math Chess and its Founder Frank Ho
SSAT Quantitative Sections Preparation Volume 1
SSAT Quantitative Sections Preparation Volume 2

If readers are interested in finding out more information about Ho Math Chess, please read the following books.

Introduction to Ho Math Chess and its Founder Frank Ho Why Buy a Ho Math Chess Learning Centre Franchise Ho Math Chess Sudoku Puzzles Sample Worksheets

www.ingramcontent.com/pod-product-compliance
Lightning Source LLC
Chambersburg PA
CBHW081502200326
41518CB00015B/2347